Environmental Science and Engineering
Subseries: Environmental Science

Series Editors: R. Allan • U. Förstner • W. Salomons

A. Hjort-af-Ornas

Turning Hydropower Social

Where Global Sustainability Conventions Matter

 Springer

Anders Hjort-af-Ornas
Linköping University
SE-581 83 Linköping
Sweden
andershiort2004@yahoo.com

ISBN: 978-3-540-74453-5   e-ISBN: 978-3-540-74454-2

Environmental Science and Engineering ISSN: 1863-5520

Library of Congress Control Number: 2007938321

Cover Design: deblik, Berlin

Printed on acid-free paper

9 8 7 6 5 4 3 2 1

springer.com

Preface

This book concerns footprints of the international environmental conventions in action. Hydropower projects have been selected as test case. The study is based on participatory research into a number of cases where implementation needs to meet new data requirements made by policy makers. It is not about findings on hydropower in studies, whether consultancies or other kind. Results in the various reports are accordingly not presented here other than to sometimes illustrate how they reflect on sustainable development. The locations of projects are not even given by maps since the purpose is to see how response to changes in environmental policy may give rise to changes in methodologies. The main issue is abstract in this sense, with focus on how sustainable development may be implemented, using hydropower as an example.

The global environmental conventions are seen in this study as drivers and the activities reflected on are just that; reflections that seek to respond and lift policy into implementation style. The purpose when making references to a specific set of studies is not to look for their results but for design; how experience is brought into policy formulation to supervise the evaluator in turn to respond with new types of information for project implementation.

The study is requested by Sida, the Swedish International Development Agency. It concerns effects on hydropower planning in developing countries from global conventions (UNCED 1992; WSSD 2002). The policy context is related to WCD, but focus is on implementation. The attention in this book is predominantly on such projects in the hydropower sector where Sida has been committed, and where I have personal experience. I am particularly grateful for the Sida support, through its Help Desk at SLU, to this study and for comments on drafts by four persons, Uno Svedin, Formas, and Karin Isaksson, SLU, commented on structure. TiiaRiitta Granfelt, ENS, gave detailed reflections on texts relating to hydropower and SIA in Vietnam. Ngoc Pham Thi Bich, VIWRRC, provided her insight in participatory IWRM. My intention with the study is to address some major trends in how social and poverty issues have been dealt with over the last decade, in order to document the new approaches in feasibility studies necessary to follow upgraded environmental policy. With this study reporting back is made to Sida and to the consultancy companies who have kindly agreed with the production of this report.[1]

Poverty reduction is a key issue for the social evaluation in many of the listed projects. This is in line with a demand that hydropower shall contribute to rural development as a sideline to energy production. Rural electrification is one important element in mitigation for hydropower construction. From the point of view of reducing poverty, a hydropower project is a potentially beneficial opportunity. But most attention is on negative impacts on environment and society. A better understanding than in the past is a key element in new policy. The question is how added information is generated. The study deals with how information is formed and managed. A main point is that this must be done through the involvement of various stakeholders, also poor people. An interaction process can be opened in this way, so that required information can be accessed both for design and impact evaluation. The introduction of a hydropower project means an unusual resource flow into a rural area. There is great potential to speed up development processes. The rural – urban migration is the most conspicuous example. The growth of rural or small towns is an alternative to urban slums. If properly supplied with infrastructure, notably rural electrification, this development can be a stabilizing feature in between rural and urban poverty. At the rural end of the scale for hydropower development, two social issues dominate; resettlement consequences and downstream effects. The track record for hydropower projects from both is bad, and issues that help to better specify key concerns are raised in this study.

The study brings out experience of change in environmental policy. It is not prescriptive but extends into cases where the new issues have been addressed. The concern is how this has been done; issues formulations and methodology approach to deal with them. The key question of the study is how policy may mitigate worst effects for the poor; and also how a specific hydropower project is linked into a development process.[2]

Anders Hiort-af-Ornas
Tema, Linkopings universitet

---

[1] My input has been sub-contracted to several consulting firms; HydroConsult, Orgut, Statkraft-Grøner, SWECO, and SwedPower. On the donor side Sida, and beyond also Norad and RNE (through ADB). The projects I have participated in, and benefited from, are Epupa Feasibility Study, Angola and Namibia (Norplan 1996); La Sirena Feasibility Study, Nicaragua (Instituto Nicaraguense de Energia et al. 1993 and 1994); National Hydropower Survey, Nicaragua; Minimum flow modeling Rio Savegre, Costa Rica (HydroConsult 2006); Song Hinh Multipurpose Study, Vietnam (SWECO 1998); National Hydropower Plan 1 and 2, Vietnam (SWECO 2001, 2004); Rural Electrification Study, Lesotho (Orgut 2002); Song Bung 4 Hydropower Project (SWECO 2006); and Small Town SMEs Isiolo, Kenya (own research project 2003).

[2] Since this study is an assessment of experience from several projects on hydropower, IWRM and rural development, I have benefited from learning processes with many colleagues and organizations. The latter are given specific reference in connection with the respective chapters below.

Contents

Acronyms

2RRBSP, Second Red River Basin Sector Project
ADB, Asian Development Bank
CPO, Central Planning Office, MARD, Vietnam
CPRGS, Comprehensive Poverty Reduction and Growth Strategy, (Vietnam)
DPC, District People's Committee (Vietnam)
EIA, environmental impact assessment
EVN, Electricity of Vietnam
Formas, the Swedish Research Council for Environment, Agricultural Sciences and Spatial Planning
GNP, Gross National Product
IWRM, Integrated Water Resource Management
KES, Kenya Shilling (71.78 to the USD)
M; LSL, Lesotho Loti, plural Maloti (6.18M to the USD)
MARD, Ministry of Agriculture and Rural Development, Vietnam
MDG, Millennium Development Goals
NGO, Non Governmental Organization
NHP, National Hydropower Plan Study (Vietnam)
Norad, Norwegian Agency for Development Cooperation
PAP, Project Affected People
PPA, Participatory Poverty Assessment
PPC, Province People's Committee (Vietnam)
R&D, Research and Development
RNE, Royal Netherlands Embassy, Vietnam
SA, Strategic Assessment
SADC, Southern African Development Community
SIA, social impact assessment
Sida, Swedish Agency for Development Cooperation
SME, Small and Medium sized Enterprise
SPM, Stakeholder Participatory Model
STD, Sexually Transmitted Disease
SUAS, Swedish University for the Agricultural Sciences

SWOT, Strength, weakness, opportunities, thresholds (methodology)
UNCCD, United Nations Convention to Combat Desertification
UNCED, United Nations Convention on Environment and Development
VIWRR, Vietnam Institute for Water Resources Research
VND, Vietnamese Dong (15,870 to the USD)
WB, World Bank
WCD, World Commission on Dams
WS&S, Water Supply and Sanitation
WSSD, World Summit on Sustainable Development
WWF, World Wildlife Foundation

List of Figures

Relating Hydropower Planning to Global Conventions

This study concerns how a technically and economically oriented planning and implementation policy tradition adapts to a new global policy on environment and development. It is empirically based on experiences from a number of projects during 1995–2005. The new requirements that a planning culture faces come from politically specified criteria for sustainable development. Hydropower planning and implementation is selected as a probe into the impact of environmental conventions on large infrastructure project implementation.

A number of considerations must be made that are new to both industry and administrations. Not only are technical and economic considerations to be sustainable, but also environmental and social issues are lifted to the fore. Issues that in the past have been treated as marginal in the technically-oriented planning process must now be moved towards its centre. Traditional sector knowledge becomes challenged. New approaches emerge as a consequence. The present study describes how this process appears in the field in the hydropower sector with regards to social issues. The study is based on first-hand experience from implementation of a number of projects. They have appeared over a ten year period. There is a build-up of experience towards better compliance with the conventions. This follows on very heavy pressure towards reforming the hydropower sector. The concern is whether there will be sufficient long-lasting effects. Changed attitudes and a new hydropower development culture are a necessity for these to remain.

Shifts in Development Paradigm

The projects selected for the current study cover more than a decade from 1993 to 2006. This has been a period when new national policies have been reshaped in line with new global environmental conventions. The study shows how issues are lifted from being contextual to forming an active part of policy formation. What used to be a technical water sub-sector in planning is becoming integrated into sustainable development planning. Sustainable development as a paradigm is established. The

1

question in the current study is how this change in outlook can have an impact on project implementation: What are the new issues, and how is information being generated?

The acceptance of newly upgraded issues for hydropower implementation can vary among project actors. There is normally a technical bias in the hydropower sector. In this tradition the generic requirements of poverty reduction and stakeholder participation can be hard to accept. Capacity building and changed attitudes will have to build on experience. The study looks at this. Information that is new to the hydropower sector "company culture" must be provided, and it accounts for the experience from solving this in the selected projects. This means the development of supplementary methods – for issue identification, data gathering and information generation.

Among the upgraded social concerns that relate specifically to hydropower planning are downstream effects, resettlement, poverty reduction, regional development, social safeguarding and market development. Most of these issues have gained increasing attention at project level also for other reasons than political conventions (UNCED and WSSD). There is an increased global attention to environmental issues. They land harder on the international agenda. Focus is shifted from national to regional sustainable development. The pressure for sustainable development has come from public opinion, NGOs and political leadership.

Hydropower is a telling case. It involves international rivers, it is crucial in development, it is large-scale infrastructure, and it has provoked much international attention. All this forms the justification for locating the study of social consequences in this sector. Sustainable development adds the listed social issues to those of technology and economy in hydropower. In the past these have often been treated as contextual. They have called for attention, but only been given it in the form of project impacts on social life. Focus has been on the negative consequences that must be mitigated. The typical social account has had the form of statistics provided for calculation of compensation. Now, stakeholders with direct and indirect project interest shall involve in a process of regional development. Hydropower projects need to be multipurpose and fit with regional planning. Stakeholders' role shift from study objects to actors who defend their interests in a responsible way. They play active parts in hydropower planning.

An attitude of "business as usual" from technically branded consultancy firms and line ministries is no longer possible. Instead, a change in development standard has been required also for all infrastructure projects, also the hydropower sub-sector. Such change comprises concerns both what issues are addressed in impact assessments and how to set up increased stakeholder involvement. The study shows a process of technocratic dominance gradually giving way. Social and poverty issues become incorporated into a problem world that used to be dominated by technical and production economic thinking. As a result, the technical feasibility requirements for project implementations remain, but new ones are added.

Global environmentally oriented conventions contribute to this process of paradigm shift. Two of them set the scene for the current study. The United Nations Convention on Environment and Development (UNCED 1992) and the World

Summit on Sustainable Development (WSSD 2002) must be applied for hydropower to specify how social and ecological aspects shall add to economic ones in order for projects to be sustainable. These two conventions, along with water sector (dams) interpretations of their implications in the form of the World Commission on Dams (WCD 2000), synthesize and give major thrusts for the needed changes in project operations. The special focus on doing away with extreme poverty by the Millennium Development Goals, MDG (2000) underscores the social side. This is a recommendation under UN auspices that defines international and national targets for poverty reduction. It calls for reduction in the number of people below a defined poverty line by half or better until year 2015. Eight related and specific goals have been set up. All development projects must address the poverty reduction goal by specifying their contribution, also hydropower projects.

In order to see how the change due to UNCED, WSSD and WCD has evolved over time in hydropower planning, the study investigates a number of projects in some detail. It makes use of them as extended cases into how key issues derived from the environmental conventions are addressed. The study focuses on the social issues under sustainable development. It shows whether, and how, the social domain has become more central also for technical project design. One key policy requirement is that stakeholders need to involve for the sustainability of any water sub-sector project. In this respect, the study illuminates central features of the change process that stakeholders must face. Government institutions, consultancy firms and NGOs, among others, need not only to make their own assessments but must also interact with each other and with affected populations towards an integrated approach. The importance of frequent interaction is underscored as a way to establish a dialogue process. A stakeholder process for project implementation can be shaped into a platform for its sustainability. Project implementation will take place within a dialogue frame without major surprises for any involved part. Data can both be generated and verified during stakeholder interaction.

The study does not go in-depth into any of the observed hydropower projects. It is confined to raise issues in them from the global environmental conventions that are of key concern for hydropower planning. These new key issues for hydropower are addressed chapter by chapter (sustainable development issues, social catchments, poverty and regional development, decentralization, and hydropower sector performance). Each chapter deals with one key concern and comments on how the required policy change has induced changes in approach. Focus is on how the newly emerged socially related issues are identified and dealt with. Table 1 lists those projects that are drawn on, it gives the respective chapters where references are made to these projects. It also specifies the issue significance of the global environmental conventions for each particular project.

During the time span of about ten years in the projects explored in this study, the impact of the global environmental conventions has been felt through the gradual emergence of new issues and consequently new methodologies needed to assess them. Two of the issues, new to classical technically-focused planning, have become the key ones. They are poverty reduction and stakeholder involvement.

Table 1 Projects utilized and their relations to UNCED, WSSD and WCD thinking

Project location	Year	Ch in this study	Issue role in the study	UNCED, WSSD and WCD significance, where relevant
Isiolo SMEs and rural electrification (Kenya)	1972 + 2002	3	Long-term effects of rural electrification on business and job opportunities	WSSD: Emphasis from 2000 on the key role for SMEs and private enterprise towards sustainable development
Rio Viejo and hydropower inventory (Nicaragua)	1994–2000	2	Rapid screening of key social issues through indicator use based on scarce material	UNCED: Sustainability sought through environmental and social safe-guarding; Before WCD & WSSD
La Sirena (Nicaragua)	1993	2	Rapid screening of key social issues through indicator use based on scarce material	UNCED: Sustainability sought through environmental and social safe-guarding; Before WCD & WSSD
Rio Savegre (Costa Rica)	2005–2007	2	Downstream effects of river regulation	UNCED: Sustainability sought through environmental and social safe-guarding; WCD: Gradually increased attention to downstream effects has been reflected in the project's focus
Epupa (Angola, Namibia)	1995–1998	1 2	Stakeholder interaction, and project evaluation by local stakeholders with international backing	UNCED: Prime attention to biodiversity issues (Red Data Handbook) and to interaction with local communities to identify alternative income sources; Before WCD and WSSD
NHP Study, Stage 1 (Vietnam)	1999–2001	1	Development of an assessment methodology combining quality and quantity indicator values	UNCED: Prime attention to biodiversity issues (Red Data Handbook) and to interaction with local communities to identify alternative income sources. Partly successful efforts to

Table 1 continued

Project location	Year	Ch in this study	Issue role in the study	UNCED, WSSD and WCD significance, where relevant
				interact with local stakeholders; Before WCD & WSSD, but the assessments carried out in similar spirit to that of the WCD
Song Hinh Multipurpose Training Project (Vietnam)	2002–2003	1 4	Involving hydropower station personnel in re-settlement assessments	UNCED: Training program with the purpose to build capacity among local staff and to build bridges between technical and local culture systems; WCD: Strong emphasis on participation and capacity building; Before WSSD
Three rural electrification projects (Lesotho, Botswana, Ghana)	2002	3 4	Look into how rural electrification may mitigate poverty	UNCED: Focus on alternative energy production among the poor;
				Not hydropower (WCD); WSSD: The poverty perspective and the role of SMEs, especially female-headed, given prime attention
NHP Study, Stage 2 (Vietnam)	2004–2006	1 2 3 5	Potential and constraints in comprehensive stakeholder participation	UNCED: Prime attention to biodiversity issues (Red Data Handbook) and to interaction with local communities to identify alternative income sources. Considerably more successful efforts to interact with local stakeholders than in Stage I;

Table 1 continued

Project location	Year	Ch in this study	Issue role in the study	UNCED, WSSD and WCD significance, where relevant
				WCD: Not compulsory since this is a follow-up of NHP Study 1, but attention to the recommendations to great extent; also noting the good congruence in principal approach; WSSD: Considered through the attention to regional development and through the emphasis on poverty reduction
Song Bung 4 Resettlement plan (Vietnam)	2006	5	Connections between policy, design and implementation	UNCED & WSSD & WCD: One basis in project design for participation and consultations
2RRBSP Part A (Vietnam)	2003–2005	1	The significance of an integrated approach when dealing with poverty reduction	UNCED: A key document for the integrated design;
				WCD: Not hydropower (but IWRM and dams); WSSD: In principle, but not with full backing; focus on poverty reduction and the use of investments in the water sectors for this purpose

Adding to these are the conditioning considerations of social safe-guarding, all along with technical feasibility.[1] The attention given to stakeholder involvement seeks good governance and a rights approach in development. Stakeholders must play a significant role for sustainable development in the new paradigm. Without

---

[1] The concept of "social safeguarding" has been introduced by the development banks. The expression means that undue harm to people and their environment must be prevented / mitigated. Such policy is taken into guidelines in all stages of projects (identification, design, planning and implementation). The World Bank for example, presents its use of the term in internet, http://go.worldbank.org/WTA1ODE7T0.

their involvement, with special competence and with special interest, projects run the risk to miss especially the MDG (Millennium Development Goals).

Stakeholders must be involved and interact across various categories for information sharing. Suitable information access is necessary for all stakeholders. Transparency is a must for all stakeholder categories. These are for hydropower at least the affected local populations and their institutions, administrations at various levels, governmental decision-makers, project personnel and financiers to mention the directly affected categories. Adding to the list are the inhabitants in affected regions or river basins. Accountability becomes an issue for all involved persons. They may represent an interest or an organization. Whatever their capacity, they need to have the sense of carrying rights and responsibilities when participating in decision-making.

A number of conclusions are drawn in the study concerning, e.g. institutional capacity and stakeholder competence (Chap. 4). In the final discussion there, the forms for stakeholder participation are addressed as one key to improved performance of hydropower planning, in line with the WCD recommendations. Other key conclusions concern how social aspects of sustainable development enter hydropower project implementation and monitoring (Chap. 1). Focus is on methodology; how to do it. The ways for stakeholders to involve in a hydropower project (Chap. 2) are also addressed from a methodology viewpoint, while the regional development and poverty issues are given special attention in Chap. 3. The final Chap. 5 concludes on progress made by the hydropower sector towards a sustainable development performance with regards to relevant social issues.

Social Assessment and Sustainability Demands in Hydropower Projects

Exploring through the extended cases, the study assesses how social consequences have been dealt with in hydropower projects against the background of environmental conventions. It draws on the author's experience from identifying and addressing key issues for the social/poverty side of sustainable development during the study period. The target is to see how these issues have been tackled in project implementation and in hydropower planning. A number of developing countries are included. The reason to go into project operation is in order to provide first-hand insight into how the upgraded key socially-related issues have been dealt with. Particular ones are resettlement, regional development and socio-cultural identity.

Attention towards these particular aspects is because they are central in mitigation, but also because they can be corner-stones in drawing on a hydropower project for broader development than energy production. By building on concrete project experiences the study seeks to document an early process of a potential shift towards "thinking sustainability" in hydropower planning. This goes for national planning and also includes the local and regional consequences for development and poverty reduction. Since the location of most hydropower projects is in mountainous and poor areas, there is often a strong ethnic prefix added to poverty aspects in mitigation.

The lessons looked for in the studied projects concern how the key social issues are addressed. The establishment of additional social information in a hydropower project is today more demanding than in the past. Until the additional time constraint for this is fully respected in planning, information generation will be done under pressure. The new methods developed have to adapt to given time and budget constraints, and still provide projects with key information in an effective way. The effect is that methods have been developed that build on fast and targeted screening techniques. There can be no ambition to reach full information coverage in the different stages of a hydropower project. But attended issues are carefully selected to represent a full and complex social situation. Stakeholder involvement in such selection processes is decisive to maintain perspective and focus.

The potential but also the constraints when applying screening techniques under tight budget restrictions are critical issues: There is a potential in terms of cost efficiency for project budget through selective focus rather than total coverage of social issues. This can be attractive from project management viewpoint. But furthermore, there is also a potential to help focusing data and information, moving a social study more towards process understanding and less of generating descriptive statistics.

There are limits to this trend towards screening techniques. The applied screening methods presented in the current study are project specific and hardly suitable for generalizations. The screening methodologies are suitable in early planning stages only, when focus and perspectives are set. Even though output gives a stringent impression since results are highly formalized, it must be accompanied by a reminder about its relative value.

The greatest single value may be in the process results are generated. The study suggests that several of the social screening techniques employed in the studied projects can bridge and open for stakeholder involvement in the social parts of a hydropower project. The result is an interaction process approach. Social information is produced in a way that can address also sustainable development issues, broader than those raised for technical constructions in project implementation. When stakeholders become involved both in design and application of a hydropower project they prove to raise not least issues of poverty reduction and regional development. This methodology development towards a design of rapid techniques then becomes of interest also beyond hydropower for sustainable development.

A key concern remains with their appropriateness for leading over into more penetrating social assessments as needed. This concern is all the more important since experience accounted for in this study shows that findings are easily overinterpreted even when warning flags are hoisted about their limits. Results about "the soft issues" of social development are easily lifted to a political agenda surrounding hydropower projects. Here, too far-reaching interpretations of results seem to be a common phenomenon. Again, this proves that the sector is in the process of adapting to the new convention demands. It needs support and capacity building for upgraded performance.

The pace for assessing social impacts in project planning is invariably high. Only the most prominent issues can be highlighted. Yet the complexity of social impacts is significant. Techniques are therefore needed that can operate through indicators and has capacity to focus on crucial issues, such as resettlement, ethnicity, poverty, regional development (upstream and downstream), migration and market access. This study illustrates through project examples how methods are applied to deal with such key social topics for hydropower project development.

A related concern, following the increased importance ascribed to social issues, poverty reduction included, is how sufficient key information can be generated during a project life. It is crucial to compile "right data" in order to achieve a balanced information base for understanding social change as caused by a hydropower project. The knowledge base has to be sufficient and suitable for strategic decisions leading to an environmentally and socially sustainable development.

Departing from participatory observations in several countries in the South, the study considers how to further an understanding of social consequences at a level that is relevant for sustainable development: How does one go about designing a way to build a dynamic knowledge base about hydropower construction, open for additional and supplementary information? There is an almost universal location for these large-scale and capital intensive but fairly short-term development projects in the midst of the poor, often among ethnic minorities. This is due to the challenge to minimize environmental consequences by placing dams high up in mountainous areas, which typically are inhibited by marginalized or minority populations. From a sustainable development point of view hydropower projects would be a suitable vehicle to reach crucial social targets.

A shift to increasingly involve stakeholders, with local but also broader regional interest, can be documented for the studied projects. Their profile has changed over the years studied. The trend is that social groups no longer are confined to form study objects and part of background data in a hydropower project implementation. Instead they also have an increasing role as actors in a selection process for key information. This is a new role expected from stakeholders in data formation. The cases suggest generally that once stakeholders sense that their work input is asked for, their local knowledge is highly valuable at project level.

In some of the studied instances doubt is expressed over the potential in the stakeholders involvement approach. The concern is that they must by definition have a limited world view in the sense that hydropower project consequences can not be visualized. This comment leads to a further question of how the information attained is used, if at all, during project implementation. The new thinking is that the value of information generated from stakeholder processes is limited (to mobilization, capacity building, and some data generation) unless a dialogue can be established with technical and environment specialists. If, on the other hand, a continuous interaction can be facilitated between the different knowledge systems, the local and the scientific, the effect will immediately be highly significant for sustainable development.

This kind of new thinking in hydropower to mobilize stakeholders at various levels, inspired by the necessity in the global environmental conventions, has led to many good intentions. Nevertheless, there is a critique that not enough have happened after the WCD recommendations were compiled. The study suggests that this is mostly due to slow institutional change inside the systems that implement projects. A further change includes attitudes and proper perceptions about addressing social issues in a sustainable way. This can be labeled to be a profound paradigmatic change within the traditional hydropower culture.

From Social Impact Assessments to Stakeholders' Involvement

Hydropower projects have far-reaching social consequences. People of many walks in life are affected, for good or for bad. Beyond the obvious purpose to generate energy, hydropower projects are nowadays as a rule therefore designed to be multi-purpose. Benefits include improvements across water sectors, such as flood protection downstream, irrigation, domestic water supply, fishing and transport; whatever the terrain allows. Drawbacks in turn with hydropower projects are many and much highlighted in international environmental debate. Widely recorded negative consequences for human populations apart from the obvious and far-reaching resettlement-related impacts include dramatically reduced fish resources, loss of fertile cultivation areas due to the reservoirs but also riverbank erosion, irregular and insufficient water releases downstream, etc. Quite obviously the significance of social issues has been vastly underestimated in past hydropower development. Project outcome has in many places been tragic, even traumatic, for individuals and groups who have had to suffer consequences without sufficient mitigation. At the same time others have harvested the fruits of social development connected to improved access to electricity, infrastructure and new economic opportunities. Such increased imbalances lay the foundation for non-sustainable development. One question is to what extent a change in paradigm can draw on the negative lessons by introducing a new style of project implementation, where a hydropower project becomes linked into regional or rural development.

The most basic social issues for hydropower projects are normally made up of resettlement and related population movements, plus downstream effects of river regulations. For the sake of highlighting the sustainability in hydropower project development, the current study links resettlement and regional development processes. The economic and political attention an area is receiving for some period of time due to a hydropower project needs to be integrated with existing regional development plans. A key issue for the regional planner is in this situation how best to balance the detrimental and beneficial consequences from an imposed hydropower project. A strive for least socio-cultural damage should combine with an active rural development. Therefore, all hydropower projects need to be placed into the context of a broader sustainable development than merely electricity production. Needless to say, the growing stakeholder involvement is a positive trend for social impacts to up into broader development. The lesson is that any hydropower project needs to be linked to the reality where it appears. This is best done by connecting with regional development socio-economic plans whenever feasible.

The question of development requirements, such as hydropower to meet energy demands, leads to a fundamental methodology concern over hydropower studies from the sustainable development viewpoint. When addressing issues and solutions in a multi-stakeholder set-up there is a need for building up a suitable data base of voluminous social data. The days of simply pulling together some kind of a set of standardized information must be gone. Instead an integrated assessment process approach in interaction with stakeholders is required. It should include the screening of current and potential situations, carefully scrutinizing the most effective indicators of likely development scenarios. The focused methods on social processes must be designed with capacity to warn against serious problems if they arise, so that more in-depth, for a project budget also more expensive, social studies are called for. The study provides examples of different approaches in countries like Angola, Namibia, Nicaragua and Vietnam, highlighting what is practicable and ethical. One issue for regional development is how relevant information can be generated so that it links into broader data bases: Linking up is a necessity for rural development. This can be considered thoroughly in project design, such as for the NHP Study, but is hard to see in implementation and monitoring. Further integration is needed but remains an ideal when the requirement for the study design is to minimize costs in the project budget without considering the loads of other budgets. Focus remains on minimizing negative social consequences rather than contributing to regional social development.

Social studies in the selected cases have provided data and knowledge for hydropower project planning and design, primarily by focusing on potentially negative consequences. Planners pay some attention to positive effects as well, but, as was just mentioned, their prime goal so far has been to minimize suffering through proper safe-guarding of human rights on the basis that no one shall suffer worse situations after than before a project. With the emphasis on combining rural development and sustainable development, an opportunity opens up to go one step further and see a hydropower project as a resource injection for an area, contributing to indirect and positive effects on regional development. One example is the ways in which hydropower production can link with upgraded rural electrification. The consequences may be promising for private enterprises and for creating jobs, apart from providing improved life quality.

Two possible methodological approaches for studying the spin-off effects from hydropower projects on rural development are presented in the study. One is to concentrate on a specific situation and make a diagnosis of what new niches may appear and among which social groups this may happen. Notably women and poor people may benefit. This is the common practice since project planning does not have sufficient time depth to follow social change. Data series are only available from administrative units (such as communes or municipalities) and not from project affected areas. This is an issue addressed in Chap. 2, where the notion "social catchment" is applied for such areas.

The other approach is to follow a local process over a long time. This will of course be a necessity for sustainable development assessments in the future. Studies on hydropower impacts need to be designed to prepare baselines. A stakeholder involvement process should a viable building block for this. In the current study one small case is mentioned; that from Kenya and the goal of rural electrification after 30 years to see long-term effects in terms of economic development and the creation of new jobs in an urban centre. Projects in Kenya and Lesotho are used as examples and also linked with some further experiences from rural electrification in other countries (Botswana and Ghana) in Chaps. 3 and 4. Assessments over time, such as in the Kenyan case, are rare. Yet, they are very much needed for understanding sustainable development trends. The long-term monitoring of hydropower projects is upcoming and promising for sustainable development. From regional planning point of view hydropower projects form one water sub-sector only. The build-up of stakeholders participating in project processes can stabilize regional and national planning. Judging from the projects studied, however, the political issue of giving weight to the result is light-weight. One striking example among them is how stakeholders' priorities (and formulations of project proposals) of water sub-sectors in the Vietnamese case of the Second Red River Basin Sector Project (2RRBSP) did not win the donor's attention. The positive side is that the administration then interfered and invested over regular budget.[2] This may be seen as a positive effect of the comprehensive decentralization political process in Vietnam. The experience can illustrate how stakeholders may apply their rights and duties by involving in sub-project planning to a hydropower project.

The Contours of Stakeholder Driven Implementation

Poverty, notably among ethnic minority groups, is a key issue in several current hydropower project sites. The concern to reduce poverty is often connected with a rural electrification issue to improve livelihoods through an upgraded access to electricity. Also for regional development more generally, investment in infrastructure project such as hydropower development must be made in a socio-culturally sustainable manner for the local populations. If this is successful local areas can benefit from resource flows that probably have not been seen in the past. The methodology issue to assess such broad processes for poverty reduction is not accounted for in full in the current study. But an approach is indicated. The study follows up the stakeholder analysis approach from UNCED (1992).

One social category of people concerned with hydropower is naturally formed by those who will actually be directly impacted. These are the Project Affected People, PAP. The question is to what extent they form a target group. They can be specified in socio-economic terms for assessing regional development impacts. The direct PAP are all identified in project preparations. The indirectly and partly PAP have to

---

[2] The proposal from stakeholders to ADB (2RRBSP Part A, Activity 3, Phase 2) for the project investment plan for the Uplands was turned down by the project but supported by the authorities. The proposal is currently (2007) implemented over regular budget (reference CPO).

be estimated, judging from the study cases in an approximate manner. This has to be done for mitigation purposes.

Taken together the various PAP social categories usually form a significant part of the vulnerable in a river basin population. The issue of sustainable development in regular rural development planning might therefore link up closer to the hydropower studies when such planning comes to an area. Commonly an urbanization process follows a project: Properly planned, with electricity supply included, and other infrastructure in place, migration to rural centers can counter the dramatic emergence of mega-cities, and reinforce a policy strategy for rural development. Experiences from Lesotho and Vietnam are quoted in Chap. 3.

With the requirements of sustainable development, many consequences from hydropower plants need to be monitored. Social and environmental assessments have to be linked into regional planning as just indicated. This will open up for longer time series for impact assessment and broaden the category of indirectly affected people also beyond hydropower. Experience is given in Chaps. 1 and 4 from creation of robust models both for monitoring and for alarming if human situations deteriorate. Stakeholders in hydropower implementation are involved also in this creation. Through such involvement they will more readily be able to take over the monitoring after suitable training. Two examples from Vietnam are documented. In one case a technique is presented to integrate technology and local competence in the hands of affected people. The target is to monitor changes in livelihoods at village level (2RRBSP). In another case local hydropower station staff and resettlees interact to shape warning models, including selecting model indicators. These indicators have sufficient pregnancy to express needs for more in-depth studies into identified key issues in resettlement situations.

Such participation by local staff and resettlees leads over to inquire who the stakeholders are. Project operations in Angola, Namibia and Vietnam are quoted. According to conventional project thinking stakeholders form part of the project issues, so that they appear as a heading in EIA and SIA. Planners have accordingly carried out stakeholder analysis. But stakeholders may also be seen as forming a driving force rather than an impacted category. In this view stakeholders' perceptions of a project needs to be incorporated also in hydropower planning, so that local communities are involved in project decision-making. Ideally, participation in information and decision-making should be by elected representatives of various affected categories. This is normally not politically possible. The second best solution is to mobilize a specially selected set of stakeholders with representative profiles. The discussion in Chap. 4 is based on an analysis of such a process, and what competence levels are applied for decision-making. The specific concern for the current study is if the chosen stakeholders have sufficient political power to influence decision-making in project implementation.

The Significance of Social Information in Hydropower Development

In the end, the study comes back to the question of how the relevant social information is used in hydropower projects construction and operation phases. There is much debate about policy and principles in R&D (research and development) circles. In reality, however, during the project implementation phase much of the sustainability talk looses ground to conventional compensation and cost/benefit approach. The current study therefore seeks to be empirically based and give insights into a number of implementation cases.

Tradition and company culture within hydropower industry and finance as well as in administration make up a global community with common values and policy features. A comprehensive change in institutional capacity is called for in order to promote a sustainable development paradigm. The policy consequences from agreed global conventions need to penetrate more efficiently into the world of project implementing bodies, also including the consultancy sector. There is a growing insight into how demanding the change in policy actually is. During some of the projects studied individual consultants have shifted around with professional "hats". An agronomist can take a role of an ethnic minority specialist and a water engineer can act as a small enterprise development expert in order to fill up roles for a contract instead of providing competence. A new ethical code for hydropower project implementation is definitely called for against this background. But the transformation into a new world order, based on sustainable development, appears to be a slow one.

The requirements for sustainable development policy implementation raised in this study are based on observation of project implementation. Some principal ones are:

A. Data and impact assessments must be compatible with regional development requirements
B. The role of, and consequences for, local populations shall not be confined to those who are directly affected by a project
C. Time horizons for impact assessment studies must expand significantly
D. Involvement of stakeholders should include an understanding of their decision-making power
E. A change in company culture is needed among involved enterprises through institutional capacity building
F. Ethical rules for consultants to establish professionalism also in the newly demanded sectors of impact analysis need to be expanded.

These are all reflections on what should be. However, the focus of this study is to document how new requirements have led to new or modified methods to respond with the required data and analyses. In the following chapters, the study will accordingly reflect upon key issues for poverty and food security (Chap. 1), data formation both for understanding and for shaping sustainable development processes (Chap. 2), regional development in the wake of hydropower projects (Chap. 3). Methodological approaches to achieve sustainability goals are documented in Chap. 4 with the experiences pointing out that both hydropower staff and local decision-makers need to cooperate with local populations in stakeholder processes. Then informed decision-making may appear in all steps, from early design to project implementation and monitoring. The study suggests that this kind of a process is possible but also wonders in Chap. 5 how to create room for it in situations under political pressure. Relevant new issues have come up on the hydropower agenda. Yet a remaining technocratic culture, especially in very large-scale projects, is blocking adjustments needed in project implementations.

Focusing on Poverty, Stakeholder Involvement and Data Formation Processes

Chap. 1 of this study concerns the basis for a pro-poor approach to hydropower design. It addresses resettlement, rural development and new demands on data formation in response to global conventions and recommendations. Data formation process and stakeholders' involvement in it accordingly become a common theme throughout the study. Stakeholders can involve in the design of assessment methodology, and in the management of information. They can also involve in the assessment itself. Such a plan may focus on the process of competence build-up for decisions in line with sustainable development thinking. However, the process also has to be rigorous with reference to hydropower development needs. Therefore, a method has been developed within the NHP Study, with the capacity to allow for comparisons between selected projects, both in the social dimensions and when combining with environmental, technical and economic issues for a number of hydropower projects. This method is accounted for and discussed in Chap. 1.

In Chap. 2 the changed requirements for stakeholder analyses and process involvement, also by Project Affected People (PAP), as requested by UNCED and WSSD are scrutinized: The situation has altered from one of dealing with PAPs in a general and descriptive way, to their involvement and targeting key issues. This way of mobilizing stakeholders has become significant to the extent that modern project design has a potential to be completely different from the conventional top–down and static engineering tradition. To some extent this is a consequence of global environmental recommendations, and the awareness raising they have brought. The study focus is on the new techniques needed, but also indirectly assessing how successful policy change can be.

Stakeholder involvement is invariably emphasized as a fundamental side to a successful development process. The study illustrates three steps: (1) The general development goal of contributing to sustainable development through poverty reduction applies also to hydropower development through a requirement of providing energy in a sustainable manner. (2) The sustainable development goal also has an implication for regional development through joint efforts between private sector, civil society, and the state (WSSD). (3) In particular for hydropower planning as addressed in Chap. 3, there are expectations that rural electrification shall have a pro-poor impact through improved production and social services. The study notes that rural electrification has no automatic take-off effect, but it needs to combine with other development incentives. Thus, electrification alone cannot contribute to sustainable development for the poor. Such observations can be made through an increased attention on regional socio-economic effects from hydropower projects.

The situations for resettled populations must be given careful attention over time in hydropower planning. If not, there is great risk that basic social sustainable development targets, with roots back as far as to the United Nations World Commission on the Environment and Development (WCED 1987) or earlier, are entirely at stake. This insight into the demands for sustainable development has become controversial for hydropower development since the affected peoples' needs for support over one or two generations become very costly. Nevertheless, the study stresses how important it is for sustainability to pay attention to these long-term effects by generating alternative income sources. In particular it indicates how monitoring of change can be developed through close stakeholder interaction to establish early warnings of undesired effects from sustainability viewpoint. A range of stakeholders must be mobilized and interact in order to contribute respective competence. Chap. 4 elaborates the selection of stakeholders; who they are, how they are recruited and what they can contribute to project design and implementation. The study notes from the hydropower case that recruitment into project planning may at present be highly selective. The issue of power and influence is touched upon; however, the study ends up in a conclusion that a representative stakeholder performance is hardly yet put into practice, in spite of international/global conventions.

The final chapter documents how the transient period of change in outlook for implementing bodies, including consultancy firms, is by no means over: The study exhibits a still ongoing process within hydropower planning. In spite of a major progress at the global level towards specification of sustainable development requirements, such as in the WCD, project implementation is still lagging behind. There is a need for a new culture of hydropower planning.

Cutting through the issues in the study chapters, two concerns are recurring: (1) the generic demand for poverty attention has to link with popular participation and (2) the need to harmonize social considerations with technical efficiency.

From policy documents it may seem that participatory processes in themselves should be the universal success tool to sustainable development. The current study arguments how false this belief can turn out to be: It is the *establishment of interaction* between stakeholders and technical expertise that is the fundament for project planning for sustainable development – participation is only the means for achieving it. Facilitation has to contain not only communication but also transfer and interpretation of technical know-how. An interaction process between the different knowledge systems is the key to sustainability. PAP (project affected people) need to interact with technical expertise in order to build competence and give feedback. Such an interaction allows better-informed decision-making where technical, socio-economic and environmental knowledge and considerations by all stakeholders can be present throughout the planning process. With this kind of a holistic approach to planning and implementation, sustainable options for long-term development can be identified from different viewpoints. The approach is transparent so that decision-makers can see the pro-s and con-s of different options. Hydropower development can then connect with sustainable development in all its dimensions; ecological, economic and social.

Chapter 1
Sustainable Development Issues in Hydropower Planning

A number of issues have appeared in the selected hydropower case studies with reference to environment/development (UNCED 1992) and to sustainable development (WSSD 2002). The thinking from environmental conventions has penetrated into action, as seen in project implementation. Departing from the World Commission on Dams report (2000), the chapter notes how similar its concerns are to those that emerged in the studied projects: Poverty reduction, resettlement issues, regional sustainable development, and stakeholder involvement both in data formation and in data assessment processes.

1.1 Project Adaptation to Sustainable Development Requirements

Ten years of methodology development – of screening and of partial mobilization of stakeholders in hydropower project design – have built an experience base about social consequences among consultants and hydropower project staff. There are many different views on the significance of stakeholder involvement, however. A significant notion made in the present study is how differently the task of making social considerations can be perceived, even after the recommendations from the World Commission of Dams (WCD). Demands on the social consultant may span from minimal fulfilling of routine requirements in checklist format, through proposing site-specific mitigation of negative consequences, to considering a hydropower project as an opportunity to reinforce regional development.

What the current study intends to achieve is an assessment of the trend to respond to UNCED and WSSD, based on experience from a number of projects. The illustrations of social studies in hydropower projects used come from first hand experience in Angola/Namibia (hydropower feasibility study of Epupa hydropower project), Lesotho (rural electrification), Kenya (small town development with electrification in Isiolo and Meru), Nicaragua (feasibility study of La Sirena hydropower project plus a comparative ranking of several hydropower options) and Vietnam (training project at Song Hinh, Ya Ly, Bac Mo hydropower stations, the comparative ranking of potential future sites in the National Hydropower Plan (NHP) Study, and an Integrated Water Resource Management study in the Red River Basin). The current study is, however, not in any way an account of consultancies done. The author has gone back to notes and unpublished reports in the various projects in order to revisit the ways in which social issues have been addressed in connection with water resources management and hydropower development.[1] The projects form the time line that can demonstrate a competence build-up in hydropower in response to sustainable development over ten years.

Hydropower projects are large scale, and consequently have potential to perform as major development engines in both local and regional settings. Consequences reach beyond environmental and social impacts. They are so significant that a project needs to be incorporated into any project planning procedure where it is located. Many proposals have been made about how to formulate linkage requirements at project levels, synthesized through the World Commission on Dams, WCD, among other initiatives.

Over the past decade a clear development trend can be seen in hydropower development. The process started with a paradigm shift from a conventional technical focus into environmental considerations, and continued into an increasing social emphasis on participatory approach and information sharing. Insight of the significance of social issues has been present and gradually turned into an independent issue, in line with the definition of sustainable development, rather than being an aspect only of Environmental Impact Assessment, EIA.

In hydropower, concepts like "social safeguarding" or "high reward/high risk" signal the investor's perspective. Calling the social dimension for "social safeguarding" and dealing with it as an aspect of EIA may, however, suggest that social issues have causal connections to environmental ones, and not to be treated as one of the three independent parameters in sustainable development (economics, environment and social) as recommended in global conventions (UNCED and WSSD). Instead of paying attention to stakeholders as important actors, they are seen as a passive category. Even if ample space is ascribed to social issues, an effect will easily be that social and cultural dimensions, like peoples' initiatives to benefit from a hydropower development, become reduced to an aspect of the environment. Or, attention may be directed solely to resettlement, minimizing the socio-cultural effects of a project on ethnicity, poverty, health and education.

In both cases the affected people become reduced to demographic figures instead of being part of a process, ascribed full respect as a dynamic resource, providing social and cultural capital for development initiatives. The consequence of this kind of thinking is to limit the complexity in social issues, in itself a necessity in rapid screening, but to do that in a way that reduces attention to the socio-cultural quality.

Sustainable development and hydropower development do not easily harmonize in spite of rhetoric among planners and developers. Different targets on the development agenda must combine, such as efficiency in energy production, poverty

---

[1] The author has been sub-contracted in the various capacities of team leader (social), resettlement, community development and ethnic minority expert.

reduction and human rights. The actors are many, both on the national scene and in the donor community, and often there are obvious communication gaps on both sides. This classic problem is found in many development sectors. In the end it reflects a constrained national administrative capacity to coordinate with donor communities, to the extent that priorities and weight in social issues are diverse and not always agreed on by different parts.

The intense international debate on hydropower dams has impacted some planning measures in a peculiar way. Technical planners find themselves torn between criticism they do not really understand, since issues raised fall outside their field of competence, and demands of upholding the economy in a project as their frame of reference in a planning process. The effect is that any new ideas connected to sustainable development easily get turned down or replaced by "business as usual" whenever possible. Old frames of reference remain through an established management culture, with traditional perception of "reasonable" cost limits for "soft" social considerations. Attacks from lobby groups against hydropower have furthermore provoked a defensive attitude from technical project management staff. The publication of the World Commission of Dams (WCD 2000) has in this respect had a constructive effect by providing concrete recommendations that open up for communication between envoys of different perspectives.

The selected cases in the current study, from Africa, Latin America and Asia, cover mainly the latter part of the 1990s and the first years into the following decade. These are the years when the WCD took shape, prior to being published until its implementation as an agenda-setter. The cases host specific struggles with methodology, transparency, stakeholder involvement and conflict resolutions in various implementation processes with rigid technical and economic preferences.

The fundamental concern in the social case studies referred to in the current study has been similar to what appears in the WCD; to reduce detrimental effects of hydropower projects, and using a logical step-wise approach. In the hydropower projects, focus is not solely on the social issues. Rather it is on demographic and natural resource data needed to calculate mitigation costs. The current study goes broader and concerns how to identify key issues. It does so by drawing on the various ways in which stakeholders have been involved (discussed in Chap. 2), from being a passive social category (even defined in demographic terms only (as in the case of La Sirena in Nicaragua) to being designers of evaluation models for resettlement studies (as in the cases of a project concerning Song Hinh, Ya Ly and Bac Mo hydropower stations in Vietnam).

1.2 The Impact of WCD on Hydropower Planning

The World Commission on Dams, WCD, has through its formation process 1998–2000 documented a global concern among its various partners over un-motivated environmental and social suffering due to imbalances in planning of large-scale hydropower projects. Until its start technical and economic considerations had been dominating, at least at project levels. The multi-stakeholder commission came up with recommendations that might not be applicable in all situations, but neverthe-less provide a philosophy that should be universal. This philosophy is summarized by the commission into five main points that also prove essential to the current study:

- Equity
- Sustainability
- Efficiency
- Participatory decision-making
- Accountability.

These are, according to the WCD, the basic values to be associated with hydropower dam constructions. Since they are values, they are not much contested until their implications at project level become clear and have concrete consequences in terms of planning and budgeting. This is why the current study is undertaken – given the gap between intention and reality, the hydropower sector should provide a good case for understanding how sustainable development recommendations are received in real life.

Implementation is also a major issue in a follow-up study five years after the WCD by WWF (2005). This study provides a global comment on hydropower de-velopment based on a few project illustrations. The WWF report concludes that im-provements have taken place, but not as much as should be. WWF therefore seeks to put stress on governments and dam builders:

> With increasing pressure to develop new dam projects, in particular in developing countries, now is the time to ensure a more systematic implementation of the WCD's recommenda-tions. They are as important for reducing the extensive social and environmental damage caused by dams today as they were five years ago. WWF is convinced that applying the WCD's framework, adapted to individual country's situations, will result in better decision-making and projects that have less impact. The world's ailing rivers and the communities that depend on them face a bleak future without prompt action.(WWF 2005:14)

The degree of detailed application of the WCD is also debated in the countries visited by means of the current study. The five bullets above harmonize with the experiences accounted for here of poverty (*equity*) (especially Sects. 2.3 and 3.1) and sustainable development (*sustainability*), as well as the application of methods for rapid screening as a means to concentrate resources on where attention is most needed (*efficiency*) (Sect. 4.1), stakeholders' participation (*participatory decision-making*) (Sect. 4.2) and degrees to which they represent all interested partners by informing in a transparent way (*accountability*) (Chap. 5). The fact that even the order of significance of the issues in the current study is parallel to that of the WCD is suggestive. Similar problems make themselves known out of real needs, and not by design, whether at project or global level assessments.

Despite the clear principles in the WCD, critical or radical words are still very much called for. If there is a battle between perspectives, as many suggest, then all good forces need to join, in order to implement the global environmental conven-tions, now latest out of the World Summit for Sustainable Development (WSSD 2002), so that balance is maintained between livelihood needs, and environmental and economic considerations. There are many attitudes that need to be changed accordingly. Donors and governments deal with several related issues but with different pre-set agendas. The same goes for some environmental groups, with conservationist priorities. One generic methodology issue in the present study is to question if there are balance points between different interests, and if so, how they should be found beyond the principles listed above.

The specific concern in the current study is a socio-cultural one, but more specifically concentrating on what kind of a methodology is required to initiate a process whereby governments, dam builders and conservationists come around and focus even more on social processes in the context of sustainable development. For instance, should hydropower energy production be seen as an international interest? Should benefit-cost ratio (BCR) calculations be replaced by other indicators of the economic side to sustainability? Should there be elections to positions as stakeholder participants in the different phases of a hydropower project?

Whichever way the debate over implications of the WCD is going to move, there will be economic considerations in one way or another for the decision-makers. Stakeholders have to formulate if they can or cannot afford an action. The way in which the assessment techniques in the current study are designed (see Chap. 2 and Sect. 4.2) is aimed to make the options, in economic terminology the real costs, more visible.

The formalized approach that is presented in the current study, is used in the situations accounted for as an initial technique at pre-feasibility. To some extent they can also be applied at feasibility level in a project planning process, but only with severe restrictions The study warns against this; the methods developed and presented here generate results that are often applicable only to a small group of projects. Data format may look general, though, and it could be tempting to draw too much on conclusions. The best example in this study is from the NHP Study where social and environmental aspects are scored and ranked. This is done on a relative scale for a group of hydropower projects, and does not express whether a project is particularly harmful or harmless.

The outcome from the screening methods presented is a set-up of typical profiles for project alternatives that can be internally compared. The simplified profiles are surrounded by warning flags for special issues that must not be lost into averaging models, such as calculations of BCRs or poverty lines.[2] This technique can be a major step towards assessing sustainable development in a region, say a river basin with a cascade of potential hydropower stations. It would, of course, have to combine with environmental and hydrology balance models. Illustrations how this could be done are given in the current study from water resource projects at local community levels.

Poverty is a key issue in sustainable development. The equity criterion of the WCD demands attention in national policies in developing countries, since all apply poverty reduction strategies. This is an issue addressed also in the global politics.

---

[2] The methods discussed have been designed in the general approach through several SWECO projects and applied into key social issues for hydropower. However, applications can be made also under other circumstances.

But already when approaching socio-cultural issues, problems of contradiction and values emerge. For instance, Komives et al. (2005:55) in their study on the effects of subsidies on water and electricity for the Poor, treat poverty in relation to a poverty line based on income levels in accordance with their World Bank thinking. They develop a methodology for assessing subsidy performance, and see the lowest income quintile (possibly with margins) as representing the Poor. In contrast, Sida (2004:6) refers to the Poor as those living under poverty; but the approach is very different. Poverty in the latter view also includes social exclusion, based on e.g. gender, so that the Poor as a category is shaped both by having low income levels and being left out in terms of inclusion and empowerment. This is also a position close to what is found in Narayan et al. (2000:4–8) that gives a scenario of the poor being left alone in spite of needs; states and NGOs being ineffective, households crumbling and social fabric unraveling.

Both approaches are official donor positions. In the situations emerging from hydropower dams, the notion of equity seems to relate much closer to the Swedish definition than that of the World Bank. The point for the current study is to note how the very basic concepts can become misleading. For the case of poverty it might help to introduce both concepts; perhaps as Poverty Mode (1) and Poverty Mode (2). This can be paralleled in the methodological application of quality and quantity indicators, a technique used in both examples from Nicaragua and Vietnam in Chap. 2 below.

The addition of a quality dimension might potentially be more politically charged in the sense that it introduces pressure towards good governance and political leadership, ascribing high priority to poverty reduction. A hydropower project plan will then have to specify how it contributes to poverty reduction specifically, and to sustainable development in the broader perspective. This approach is in line with the spirit of WCD. It would provoke a change that is necessary for the sake of sustainable development. The current study concerns methods to generate an information base for such a move, combined with the concern about where initiatives for information exchange are found.

If properly applied, one useful approach is to apply indicators that can disaggregate a complex concept by highlighting key aspects for hydropower project planning. Resettlement, poverty, conflict resolution through stakeholder interactions and regional development are pointed out as key issues in the current study. Both the process towards issue identification and resolution, along with the methods to specify these issues are of significance for sustainable development.

One suggestion given in the current study is that by cautious use of such methodology models that combine quality and quantity changes, different hydropower options can be grouped into combinations of low risk/low return and high risk/high return categories. With the quality side of a hydropower project indicating political goals implicit for sustainable development, it touches on one key issue from the WCD – accountability. A project is accountable to the combined set-up of stakeholders who form a constituency where choices like low risk/low return versus high risk/high return can be penetrated. Their judgment is the basis for the quality side of an indicator in the assessment process of a hydropower project.

1.3 Key Issues for Sustainable Development

The emerging socio-economic issues following the recommended principles of the WCD need to be agreed and assessed among diverse interest holders within hydropower planning. The current study briefly refers to a number of social studies in hydropower projects, with relevance for the discussion on key issues of relevance. This assessment of issues is concentrated on both local, directly affected level, and on far-reaching (regional) effects from dams. Analysis and experiences from both levels can provide feed-back to further understanding of the multiple effects of large dams and have bearing to both the principles and practical design processes of future projects.

This understanding of multiple effects is an important element for comprehending the long-term sustainable development of the hydropower sector. The intention with the study is to demonstrate this and suggest that social studies within hydropower need to designed in such a way that they may provide a baseline for related sustainable development studies at country, regional or local levels. The philosophy of sustainable development concerns both aspects of detrimental and development potential from hydropower projects. The key issues as derived from the various hydropower projects that the study draws on are:

- Downstream effects from daily and seasonal hydropower regulations
- Resettlement problems and related issues of compensation for affected people
- Poverty reduction potential as an evaluation criterion for hydropower projects
- Social safeguarding against detrimental social effects
- Regional development through investment in a hydropower project
- Rural market development due to improved energy production and distribution.

Downstream effects are addressed in Chap. 3; early thinking assumed proper regulation and flood control, whereas reality can experience regulation geared by daily electricity consumption patterns, driven by profit thinking, again in a tradition of meeting solely classic economic and not other demands on sustainable development.

Resettlement is a key concern in the social safeguarding perspective. The design to minimize it, the approach to make it as smooth as possible, and the long-term sustainability are prime concerns (Sect. 1.4). The study addresses the last point in some detail; an approach is accounted for that has been tested at three hydropower stations in Vietnam. Local staff has developed indicators, and partly a model, that allows for local monitoring and early warnings when social situations for resettlees deteriorate. This is accounted for in Chap. 4.

Poverty reduction is today a compulsory element in development projects. It is addressed in Sect. 1.4 below and returned to in further detail in Chap. 2 in connection with *Regional development*. Effects from large-scale hydropower projects may concern water use and the extent to which a dam becomes multipurpose through integrated water resource management (IWRM). Of special interest is the effect on poverty reduction. The study includes how the water sub-sectors may relate to changes in poverty status, so that monitoring becomes possible for local authorities.

The key is to address water supply and sanitation (WS&S), relating the sector with health issues (not least water-borne diseases, of course).

Social safeguarding is an issue raised at pre-feasibility and feasibility stages. A way of identifying key problem areas and assess them is discussed below (Sect. 1.5 and Chap. 4). In order to reach closer to social reality than passive impacts, a rapid and robust methodology has been developed by combining a quality and quantity assessment of a limited number of key factors. The approach is developed further in Chap. 4. The potential of this approach is raised here on the basis of experience from Vietnam, Nicaragua and to some extent Namibia. It opens for a Participatory Social Assessment, PSA, of planned hydropower projects at an early design phase.

Market development seems most successful in connection with the emergence of small urban centers. This trend can lead to effective *rural electrification* since investment costs go down. Two cases are drawn upon. One is a study over 30 years of one township in Kenya, from the days before electrification until today. The other concerns Lesotho and the effects of rural/urban electrification on female-headed small and medium scale enterprises (SMEs). Both studies are presented in Sect. 4.1.

All the listed issues are cross-cutting in the chapters of this study. They are ordered in a sequence to follow how the role of social and poverty issues has developed in time; from providing facts and statements in a static, descriptive style, to understanding process and extrapolating likely future scenarios. Based on the experience from specific projects, the quoted key issues appear repeatedly; down-stream effects, resettlement, poverty reduction, regional development, social safe-guarding and rural market development. The study gives experience of how to tackle these issues in different situations.

The social studies in hydropower projects may reach also beyond forming baselines to future or ongoing sustainable development studies.

- In terms of sustainability aspects of hydropower Sect. 1.4.1 illustrates how priorities within a water sector can be made so that the prime aim in investment planning becomes poverty reduction. This could represent a livelihood based process option to the descriptive poverty line.
- Section 1.4.2 gives experience from how resettlement effects can be monitored over time through a combination of rapid screening and in-depth assessments.
- Experiences for regional development are given in Sect. 1.5, linking hydropower project planning into regional development plans.
- The issues of social safe-guarding and rural market development connect with the development roles of a hydropower project; whether the goal is to minimize negative social effects or, in a more ambitious way, to see a project as an asset for regional development. Such issues are touched upon in Sect. 1.6.

This list is not complete but intended to illustrate some bridges from hydropower into sustainable development considerations. The current study becomes a plea for making such connections. The initiative might have to come from the industry side. With the relative harmony between national clients, responsible for policy and reform on the one hand, and industry with responsibility for technical and financial economic viability, follow-ups of WCD could be anticipated. Some want that to be transformations from its recommendations into mandatory national development policies. The current study does not go into this issue.

1.4 Resettlement Risks and Poverty Reduction in Multipurpose Water-Sector Projects

There are two types of issues connected with practically all hydropower projects; re-settlement consequences and poverty effects. Emerging new ways to deal with these are presented in this section, drawing on two studies in Vietnam. This country exhibits an extremely good performance in terms of poverty reduction. The presented cases can therefore be of high significance also beyond the national borders.

1.4.1 Poverty Classifications During Hydropower Implementation

The assessment of poverty situations naturally depends on of how poverty is defined. The two established methods are mentioned above through the examples of Sida and the World Bank. The two approaches differ; one makes use of aggregate data measuring a life quality approach, the other defines a poverty line in income terms. The approaches give different results, depending on definition of a model indicator rather than on the experiences of suffering by the Poor.

The design of indicators is obviously a fundamental question in the current study. When addressing poverty issues in connection with hydropower projects it is not sufficient to operate solely with quantified measures at village or local administration levels. One short-coming is obvious since projects must identify vulnerable households for compensation. Projects are then down to the level of individual affected households, and the issue concerns their resource access. The poor can then be found amid plenty. The availability of food in a locality does not indicate food security (Sen 1981). Assessing food availability in aggregate terms for a whole local community, and doing it for the individual inhabiting the specific locality are widely separate things. Hydropower projects, having to identify needs and calculate mitigation costs, need to be more specific than presenting statistics that for instance a certain percentage of a local community live below a poverty line.

Resource access for project affected people is through serving institutions, including legal systems of land rights. Where rights, opportunities or power are weak, such as may be the case in hydropower implementation, the poor become the losers (cf. Narayan et al. 2000:11). In order to include the socio-cultural specific in various situations the concept Participatory Poverty Assessment (PPA) has been coined. It is an approach for understanding poverty from stakeholders' points of views and not a rigid methodology. The emphasis is on process; involvement by the affected Poor in the assessment (cf. Cambodia 2001, Kenya 1996, Lao 2001, and Pakistan 2003).

The PPAs depart from the poverty line, they distinguish the very poor as an especially vulnerable category, and they combine the static poverty line indicator with a process approach thinking for relating poverty with peoples' needs. In consequence with this, connecting to hydropower, two approaches are applied in the cases studied: The statistical one using the poverty line defined by income and consumption, or the life quality approach with a set of life quality indicators, such as social service and infrastructure, health and disease.

In Vietnam both approaches have been utilized in parallel, and the uncertainty has been demonstrated for both of them. Both have been redefined around the end of 2005. The life quality based definition caused the poverty picture to change abruptly when the indicator values (and to some extent choice of indicators) were officially adjusted on November 1, 2005. At the same time the poverty line level was also raised, its value doubled, so that the proportion of the population below it increased from 7 % to 18 % in one stroke (Vietnam News January 3, 2006). Vietnam is a success example for the Millennium Development Goals of halving the poverty from 1990 until 2015. Already now Vietnam has achieved this (a decade quicker than agreed) by halving poverty all over the country, including the uplands and mountainous areas where hydropower projects are located (cf.www.oneworld.net of May 5, 2007).

The noteworthy exception from this trend in Vietnam is that ethnic minorities remain poor. This trend is particularly significant for social studies in hydropower projects, given their locations in mountainous minority areas. Behind the observation that the Vietnamese ethnic minorities do not follow the poverty reduction tendency lies a poverty trap. Their poverty resilience has to do with how disconnected they are. This suggests for sustainable development that the MDG intention with the halving to fully do away with extreme poverty is limited thinking concerned with demographic averages only, unless also other measures are taken. Those social categories affected by hydropower development are often precisely those not included in the reduction process. The hydropower sector therefore has to address this issue for proper mitigation. From this need follows an initiative towards a design of alternative assessments of poverty, so that the vulnerable groups can be reached.

This concern to have to deal with poverty in a hands-on manner at household level for hydropower projects can be of benefit for regional sustainable development planning. The technique is robust, built on domestic water consumption. It does not include complicated water/poverty indicators. The most vulnerable, Poor households can be identified. They are the ones calling for special attention because of detrimental effects from hydropower.

In the other perspective, the positive effects for the Poor, job opportunities are often raised. Hydropower implementation, hydropower being large-scale infrastructure investment that attracts capital flow to a region, should offer opportunities in regional planning to create job opportunities for unskilled labour. This does not make it particularly interesting for sustainable development, though, since job opportunities only last during the construction phase of a project, and even so leads to much imported labour. Instead, the sustainable development should be looking for the effects from the hydropower projects' needs to build capacity among local populations to involve in implementation as stakeholders. It means a cost for capacity building for hydropower, and an opportunity for regional development to upgrade the "social capital". Effects among the poor would reach beyond the role as casual labour. By developing skills, to be negotiated into the budgets for hydropower implementation.

Working with social indicators provides poverty profiles that can be locally specific and accordingly are hard to generalise. The specific task of mitigating hydropower consequences through measures with a pro-poor profile and with an aim towards sustainable development is better carried out this way, especially when directly affected populations are small. For the purpose of the more specific demands of a Social Impact Assessment in hydropower development, a stronger emphasis on water resources can be suggested. The approach introduced below can be a platform for the design of a methodology to assess poverty impacts on the hydropower sector of integrated water resource management. The basis is that there is a correlation between domestic water use and poverty situations according to the life quality definition. This observation offers a way to identify vulnerable and poverty-struck individual households in a project area. With such an understanding in place, individual households – not a whole group – can be targeted for mitigation. The technique is an effective way to combine cost efficiency in the assessment with targeting of key issues for sustainable development, which is its social side. The Second Red River Basin Sector Project, 2RRBSP, has developed two ways in its Component 3; first using direct correlations between poverty and domestic water use, second combining domestic water and waste management with some other factors (ADB 2005b).

Even if the empirical part in the 2RRBSP is not developed specifically for hydropower development, the location of the study in the uplands of Vietnam is typical for many hydropower construction sites. And the approach can be taken one step further into sustainable development; to establish site profiles for hydropower development that include all sides to sustainable development. In the mentioned study (ADB 2005b) on poverty reduction through integrated water resource management, IWRM, a number of typical features for water-related issues, all relating to local community development were identified. The concerned villages all had their livelihoods based on irrigation agriculture. Seventeen villages, or about 350 active stakeholders, were involved in this study. In the process of interaction, stakeholders first prioritized investments into irrigation, water supply and sanitation, flood control, and rural infrastructure development (roads and electricity). A number of sub-projects related to these areas were then outlined and proposed for viability assessment. This evaluation followed the thinking in sustainable development by highlighting the three pillars: poverty, economics and environment. These were applied through five parameters at the optional project sites:

- Water availability
- Technologically suitable
- Economically viable
- Environmentally sound
- Socially prioritized.

Each of these five parameters had a design based on a few indicators. Further details are given in Sect. 2.3.1. The social/poverty dimension was elaborated by applying four sub-indicators at household level for the directly affected people: The village poverty status according to commune statistics, the availability or not of a dug well, access to a pit latrine for household members, and the ethnic minority situation. By combining these four village household issues into a profile among the directly affected people, a qualitative assessment of the social situation, such as poverty, at project level was achieved.

This approach can lead to a model in sustainable development that can be applied in many ways. For example, profiles based on the selected parameters were in the study (*ibid.*) used to select a group of sub-projects with high potential for success in achieving poverty reduction through participatory IWRM planning. Each of the five parameters had to meet minimum requirements (sufficient water, technically feasible, economically reasonable, pro-poor in orientation, and environment friendly in design). They could also be combined and stability be tested through ascribing different weights to the indicators.

The model above is the final step in a regional site selection methodology (further detailed in Sect. 2.3). Through careful stakeholder analysis and through going more and more site specific in a selection process, the design builds on a process of interaction between various significant stakeholders, technical people included, making selection and decisions wherever suitable in a stepwise implementation as follows:

- Regional or river basin stakeholders from administration and sector ministries build consensus about one priority sub-basin for investment
- All small catchments in this sub-basin are identified, and their local administrations give priority through consensus to one or two (based on rigid criteria)
- Local inhabitants (village and commune levels), local authorities (commune and district levels) and technical experts come up with comprehensive sub-project priority lists for investment in water resource projects within each of the small two small catchments
- Consensus is built around these lists. Technically impossible options are reformulated or excluded through informed consensus building.

What this example shows is an assessment process capable to integrate technology and social/poverty features through a well-organized procedure of informed decision-making. In the hydropower situation this approach may prove particularly suited since it involves stakeholders as actors, not merely as spectators, and also since it brings key sustainability issues into the project in a transparent way, balancing the traditional technical/economic thinking. For instance, all projects are nowadays in principle required to be multi-purpose, and social mitigation is a must. The concrete issues in each specific situation can be brought out through this kind of a technique. It forms a process around the structured methodology that involves stakeholders in the issue identification, which is transparent and potentially stakeholder-driven. The Poor in a community are among those mobilized by the process. Proper stakeholder involvement in assessment includes technical dialogues, so that consensus is built among stakeholders as far as possible.

1.4.2 Post-Resettlement Screening in Vietnam Central Highlands

Stakeholder involvement can go even deeper, beyond assessments of different options by involving stakeholders into formulation of an assessment model. Chap. 4 gives an example from a project with staff from three operating hydropower stations. They were involved in developing assessment models of socio-economy and health among the resettlement populations associated with their hydropower stations. This project was part of a training course over two years (2002–2003), and turned into an eye opener for the participants, predominantly technical and administrative staff.

Participants were in the start-up workshop asked to establish two lists of indicators, one for socio-economic issues and one for health and disease. Group work generated extensive lists, which were then reduced as much as possible until participants refused to cut out more variables. This process included the task of avoiding overlap between chosen parameters; that is to search for as independent indicators as possible. Participants were then requested to spend one week in the resettled villages and under supervision to gather data for the topics they had selected. This approach initiated a process of interaction between hydropower station staff and resettlees towards further reducing the indicators to a manageable number. The continued training focused on reporting, continued reduction of the number of indicators, and finally a development of an early warning model for deteriorating social situations, described in detail in Sect. 4.1.

The point to be stressed for the current study is the process, since it targeted sustainable development in the area affected by the three involved hydropower stations. The technical personnel from these were naturally not expected to actually carry out social studies. Their training in this part of the project was in interaction, which then functioned as an eye opener for them: The participants quickly broadened their outlook, realizing the plight of resettlees but also winning an insight in the fact that the social problems are not diffuse and hard to grasp as they had thought, but hands-on and can also be understood by technical staff. In this sense the project has led to a shift in problem perception for them. It has taken participants away from the technical focus and a view that social issues are a burdensome complication that requires safe-guarding, into seeing stakeholder interaction as an opportunity for referencing in the daily operations.

The lesson for participating staff as well as resettlees was the way in which stakeholding turned into a process and not an isolated consultation workshop. The engagement of the directly affected people in this case introduced a new way of looking at development; the prefix sustainable is close. The high degree of participants' involvement was at the time unusual. This has changed over the past few years, almost to the opposite. Today, there is a danger in applying stakeholder participation as a recipe for success rather than as a methodological approach (ADB 2004). Participation should of course not be an end in itself just because a handbook prescribes that. It is rather the basis for an interaction process, necessary when applying quality-oriented methods. This is described and addressed in further detail in Sect. 4.2.

1.5 Hydropower in Regional Development

In a regional perspective, a large-scale hydropower project contains an additional, dynamic development potential beyond the immediate social impacts. The case of Epupa hydropower project provides below in Sect. 1.5.1 an example of diverse ways of comprehending the anticipated regional development impacts from a project. During the feasibility study the potential spin-off for regional development consequences became a dividing political issue in Namibia, South of the border in this international river project. Parallel to this, North of the river, the Angolan debate was dominated by the positive regional development prospective in the form of improved infrastructure and a general flow of capital into the area. A similar view was held by one of the local political groupings on the Namibian side of the river, while another one was against any large-scale project, basically for environmental reasons. The political formation in the local ethnic Himba community was significant; for or against any localized, large-scale and capital-intensive development project; be it hydropower station or (as in an earlier proposal towards large-scale investment) a regional hospital. The political process shows how the regional social impact can turn into a real issue in peoples' lives, far beyond those households more or less directly affected by a technical project.

An anticipated direct regional effect is improvement in life quality through rural electrification. In the current study the power consumption issue is addressed through the cases of rural/urban settings in Kenya and Lesotho dealt with in Chap. 3. The energy consumption of different consumer categories is addressed, and long-term development trends are assessed. The experience from both countries shows that rural electrification should be seen as a long-term infrastructure investment: There is no quick fix towards massive electrification through subsidies. One feasible way seems to be to apply price regulation of connection costs, as a way to give compensation to affected communities through improved energy distribution. In economic terms such a pricing policy may not be profitable, but in a regional development perspective it opens up new avenues for unprivileged social categories such as women and the poor, and contributes to improving life quality. New activity niches can be established, and security, education and health improved.

Much attention has usually been given to the studied projects themselves, far beyond the affected local communities. These are in most cases of ethnic minority status, and the concern has been with the effects in socio-cultural terms from an introduction of a hydropower project. In some cases the ambition to uphold social safeguards has led to a protective spirit notably among environmentalists. In other cases the ambition to create regional development has led to exploitation and destabilization through an economistic hardline development spirit, not least among political leadership. Early stakeholder involvement in issues identification and solutions shows constructive results in several cases. One example from Vietnam is given below in Sects. 1.5.2 and continued into 1.6.

1.5.1 Conservation vs. Development in Namibia

With the background in several monumental failures of hydropower projects to account for social consequences in the past, there is no wonder that "social safeguards" has become a key planning concept today. This has been felt also in the projects discussed in this study. The Epupa project feasibility study on the border between Angola and Namibia in 1995–98 attracted much debate inside Namibia and internationally among environmental lobby groups. The stakeholder mobilization proved to be politically very sensitive, and was both monitored and constrained by relevant authorities. Likewise in the early phase of the NHP Study in Vietnam,[3] stakeholder and NGO involvement through workshops proved to be a sensitive matter. Such sentiments have changed, as will be shown below. The examples indicate a considerable change in attitudes and project planning cultures over only a few years. Had the projects been started today, a more profound stakeholder participation process would have been both mandatory and appreciated by planners. The experience built, and probably reinforced by the emergence of the WCD, has opened for a culture change in the world of hydropower project planners.

An internal debate between two local ethnic Himba political leaders at a stakeholder meeting on the Epupa hydropower project on the Namibian side came to touch upon a principle position concerning the regional socio-economic effects of large-scale projects: One held that society is changing, young men migrate to cities for work, etc., and there is good reason to bring large-scale projects to the rural region whenever opportunity occurs. As just mentioned in last section, the same polarization had occurred some years back when the regional hospital was proposed to be located in the area. This had been refused by the community, with similar mobilization to that of the currently discussed hydropower project. The counter argument from another leader, of the other opinion, was that a large-scale project brings so many social changes to the community that the region will suffer also culturally through immigrants and a new economy. In this view there was good reason for militant social safe-guarding, not in development project terminology but in local community views, by refusing any large-scale project.

The debate between these two local leaders exemplifies the two views on a large-scale project; either bringing social problems or bringing socio-economic development into a region. The conflict between the two perspectives could not be resolved in the stakeholder participatory process. Both views are highlighted in the following two sections, one on a potentially constructive process of resolving potential conflict at an early stage, and the other on an unresolved conflict.

The political discourse around the Epupa feasibility study in Namibia and Angola was in Namibia mainly driven by environmental conservationists. At the time great attention was paid to the project also internationally. In some of the international debates social information from the project was rejected on the basis that the project

---

[3] NHP first stage took place 1999–2001 and the second stage 2004–2006.

was destructive in itself, so that an issue discussion would mean involvement and should therefore be refused. This political side is discussed in some further detail in Chap. 5.

The stakeholder process was also dramatic. The start-up stakeholder workshop, merely aiming at providing local people with information, was done with military presence. The project proposed design, to work through representatives of local populations engaged by the project, was rejected by national authorities. Key political players furthermore complicated the situation through blunt statements about the right of the government to take initiatives regardless of local opinions. Underlying was a long-standing political rift, probably also with roots into the *apartheid* era.

Beyond a number of special issues for the Namibian situation in terms of domestic politics, the rift through the local ethnic, Himba, community cut straight through it along the lines of traditional leadership and modern leadership. These two sides could not reconcile but remained polarized and antagonistic. The traditionalists, backed by local chiefs and prominent representatives of the global environmentalist community on the one hand, stood against the local leadership and middle class, backed by the national government on the other hand. The study, including this polarization, has been registered in four films (Mathews & Chesselet 1997).

The case of the Epupa hydropower issue on the Namibian side of the border is one where the internationalization went far. Local community based issues mobilized also outside actors, so that there was a gradual shift with new influence into the stakeholder process through this influence. Western style political style lobby groups mobilized environmental groups' networks and operated not only in Namibia but internationally. The other side, in favour of a proposed hydropower project for regional development, tried to operate their political agenda through conventional national political institutions, with less success.

1.5.2 Conflict Management and Stakeholder Involvement in Project Formulation in Vietnam[4]

When the NHP Study was designed, the WCD Report (2000) was still not available. There are, however, no significant contradictions in the study in comparison with the WCD guidelines. It could even be claimed, that stakeholder participation is taken more seriously in the NHP Study than in the WCD Report through the four-layer approach presented below. The congruence in approach is no surprise, of course, since both the NHP Study as a project, and the WCD as a stakeholders' design, are outflows from the same insight about the significance of a participatory process. With the growing emphasis on sustainable development and efficient use of water resources has followed an increased expectation to regard hydropower projects not as isolated technical investments where social consequences need to be minimized,

---

[4] This text draws on a discussion for the National Hydropower Plan Study, Stage 2, where the author acted as Team Leader for the social studies.

but to place them in their capacity of being large-scale development projects with social and environmental influence, both locally and regionally.

Among key issues are stakeholder analysis and involvement, plus the ambition to minimize detrimental effects. This can be achieved by careful stakeholder involvement throughout a project. The first step in the NHP Study was to seek consensus as far as possible, on approach, issues and solutions. The series of workshops reported on in Sect. 2.2, illustrate the importance to establish a process of consensus-building around problems and solutions. In the case of the NHP Study, participation primarily involved local leadership, see the discussion on stakeholder analysis in Sect. 4.2.

A second step comes when consensus no longer is possible. It is then important that the legal system is clear and agreed on by all parties. This is one reason for the NHP Study to ascribe high priority to awareness raising and the establishment of a process whereby stakeholders build their capacity to interact in large-scale hydropower projects in the future.

The comparison of the NHP Study with the WCD Guidelines has its limits. With regards to the Strategic Assessment, SA, in WCD terminology, for instance, it must be stressed that the NHP Study is a pre-project study. The Consultants had an ambition to be far-sighted and prepare the ground for solid WCD compliance once projects are initiated, by setting up early awareness raising and broad and strategic participation of stakeholders. Strictly speaking, however, the NHP Study does not include SA, but a more simple but in-depth study into a few key issues, as discussed in Sect. 2.2.

The principles for conflict resolution are clearly addressed in the WCD Guidelines. The NHP Study shaped a basis for building consensus as early as possible in a project planning process in order to address potential conflicts before situations become infected. This is why consensus was built around potential problems at four levels; national, river basin, provincial and village. Out of these, the Provincial Workshops were implemented with about 110 key political representatives from the directly-affected districts and communes, and the River Basin Workshops by a similar number of stakeholders (around 90) but recruited differently; among district and province decision makers. The goal in the first instance was to address mitigation, and in the second to deal with regional development.

The importance of consensus-building workshops such as in the NHP Study cannot be exaggerated. The methodology has proven suitable, facilitation professional, and spin-off effects materialized in the form of mobilization and sensitization among key persons. For the NHP Study, stakeholder interaction was intended to be a first step towards permanent interplay between the project and the affected people. This interaction can focus on the multipurpose side of a hydropower project. The goal to establish participatory stakeholding shifts policy-making from control and guidance functions towards opening up for influence by the stakeholders through interaction and informed consensus building. A dynamic process is sought, involving multiple stakeholders instead of rules and regulations. As the process evolves over time, issues become more focused and new stakeholder categories can be involved, not only actors or players with high power to affect decision making, but also those with low power but direct or indirect interests at stake. The process in this way enables a setting for addressing especially long-term issues such as possible down-stream effects of water regulation or water allocation for irrigation.

Since stakeholder participation was a key issue in the NHP Study, their involvement could have been a start-up for regular consultations during further project planning and implementation, not least with poverty exposed groups. However, real life in the Vietnamese hydropower planning also tells a different story, discussed in Chap. 5. In Vietnam regular consultations by administrators have also been identified as a major issue to fulfill the national Decree 29 on grass-root empowerment. The contrast to the Namibia debate is big in spite of the presence of international lobbyist groups like the International Rivers Network in both instances. There were eventually some attacks from them on the consultancy firm but no questioning of the government policy has been forthcoming. The internationalization of the NHP Study was accordingly far less efficient. An argument for this difference has been that the centralized political structure in Vietnam is discouraging such a process. It could be, however, that the more fundamental fact that Vietnam is both well-known and a show-case of development success has hampered the international debate to reach the issue of how sustainable this development actually is.

1.6 Sustainability Indicators Applied on a Set of Hydropower Projects

Sustainable development shall be assessed in three dimensions according to the WSSD: economic, environmental and social. For hydropower projects, technical assessments must be added. With this approach in the NHP Study in Vietnam a technique for dealing with social and poverty issues was developed. It was derived from the project's goal to establish combined scores for a number of potential hydropower sites, where the technical and economic calculations are supplemented with assessments of a project's potential environmental and social impacts. This method design also allows for a continuation, outside the NHP Study intention, of ranking hydropower projects according to their impacts on poverty reduction by assessing solely the social indicators in a quantitative and a qualitative dimension.

The task for the social studies in the NHP Study was to develop a methodology for comparison of a large number of optional sites for hydropower plants. The social analysis had to be partial, focusing on key features for baseline design and future impact assessments. Particular reference was made to resettlement, livelihoods and life quality both for those directly affected (including host communities for resettlers) and for the region. The assessment had to allow that conclusions for each potential site could be fed into the other assessments of that site, from environmental, technical and economic perspectives.

As a first step the study area for each potential site was divided according to location into river basin, catchment, project and downstream areas. Secondly, 20 key issues were identified through a process of interaction with expert stakeholders,

Table 1.1 Selected key social issues for effects in different scale, the NHP Study

River basin area	Project area	Downstream area
1. People resettled	8. Directly affected people	19. Fishery
2. Host area relations	9. Indirectly affected people	20. Water access and quality
3. Ethnicity	10. School enrolment	
	11. Ethnic groups and history	
Catchment area	12. Health and disease	
4. Water-related health	13. Cultural artifacts	
5. Transport	14. Extension service	
6. Water access and rights	15. Land use	
7. Migration	16. Farm output	
	17. Market structure	
	18. Secure access to food	

mostly from the administration and technical staff. These key issues referred to the study areas in this way are shown in Table 1.1 below.

Basic data was collected for each site under each of these headings; to begin with from available secondary sources and then through targeted field studies in villages, communes and districts to be directly affected, and in provinces hosting a potential hydropower project. These studies consisted of a range of activities, such as visiting local archives in district centers, carrying out surveys at commune and district levels, and conducting specific PRA based selective studies at village level. The output was a built-up data bank with items for each potential project site. It was selective, addressing only key social issues. However, the data base also contained "warning flags", the surveyors' observations of special, particularly detrimental issues concerning vulnerability in current social life.

Out of the 20 themes 12 were selected for developing indicators for potentially detrimental impacts of a hydropower project and 6 for expected positive effects. The same indicators were used for the whole ensemble of potential project sites; see Table 1.2 below.

Table 1.2 Social detrimental and beneficial parameters, the NHP study

Social detrimental parameters	Social beneficial parameters
People resettled	Rural electrification
Host area relations	Roads
Ethnicity	Education
Water-related health	Health
Ethnic complexity	Provincial investment
Migration	Aquaculture
Partially and indirectly affected people	
Fishery	
Loss of agricultural land	
Food security	
Poverty	
Water use downstream	

The data bank was screened systematically, so that information for all sites on each of the topics was compiled. It was then assessed with reference to potential effects by a project on the communities (at the different scales) in terms of how much impact could be anticipated (quantity) and the significance of that impact for social life (quality).

For each of the selected parameters the relating features of each potential site were listed, both in terms of quantity of the potential impact and its quality, as just defined. Criteria were developed to rank them internally (two ranking procedures; quantity and quality interpretations). The output from this part was an ordinal scale for the quantity side and another ordinal scale for the quality side of the potential projects. The scales were divided into quintiles so that an equal distribution of sites was established for each indicator (times two; the quantity scale and the quality scale). Numeric values were ascribed according to the severity of a potential impact in the case of the 12 detrimental parameters, the highest value giving the worst-case scenario.

This whole procedure was repeated for the six beneficial parameters. The intervals in this case were also quintiles, and the highest score reflected the most positive anticipated effects.

In this way a profile was created for each potential hydropower site with reference to social issues. This profile could be lifted into the larger project assessment model, first by combining with a similar but less complicated assessment of environmental issues, and then with economic and technical assessments. The approach allowed a rather formalized analysis, be it in relative terms; it can say little about the objective impacts but provides a strong tool for internal ranking once the group of potential sites has been decided.

Table 1.3 gives illustrations from the procedure of two detrimental and one beneficial parameter. Magnitude denotes the quantity side of the issue; Importance represents the quality side.

It is not possible to go into the details of the sophisticated analysis of magnitude and importance for each parameter in the current presentation. The result of the analysis was a combined social score, where the numerical values were combined through adding them. The output is one combined social score, a numerical value, for each potential site. The total scores were fed into further project assessment, by combining with other analyses (the environmental, economic and technical ones).

The analysis was then taken one step further, by discussing the weight of each of the parameters. Some are more significant for economy and production in the community, others reveal more about social and cultural issues. Weighting therefore allows for a more elaborated discussion on potential consequences for livelihoods and for social life. With the weighting technique both sustainability and vulnerability can be expressed in a visual way. In the study stakeholders were asked to ascribe their subjective weight to the parameters. Fig. 1.1 shows the weighting results of the 21 studied alternative project sites according to stakeholders' perceptions, along with the results of the formal analysis where the Vulnerability and Sustainability curves have been generated by ascribing the parameters different weights, based on the appearing values from the parameter scoring described above.

Table 1.3 Three examples of social scoring, the NHP Study

Detrimental 1: People resettled
Magnitude: Relates to the number of directly affected people
Importance: Concerns resettlement distance

Very High (4)	Very large number of resettlees (x>5,000)	Resettlement outside the province
High (3)	Large number of resettlees 1,800<X≤5,000)	Resettlement within the province but also in other districts
Medium (2)	Intermediate number of resettlees (700<X≤1,800)	Resettlement within the same district
Low (1)	Small number resettlees (X≤700)	Resettlement within the affected communes
None (0)	No people to be resettled	No resettlement

Detrimental 9: Loss of agricultural land
Magnitude: Percentage of agricultural land lost in Project Area communes
Importance: Types of agricultural production

Very High (4)	16 % ≤ of agricultural land to be lost	A project appears in predominately wet-rice or cash-crop oriented situation
High (3)	Agricultural land to be lost 10 % ≤ X < 16 %	A project appears in a mixed farming situation
Medium (2)	Agricultural land to be lost 3 % ≤ X < 10 %	A project appears in an upland production, including swidden agriculture and livestock production
Low (1)	Agricultural land to be lost < 3 %	A project appears in a situation with only food crops production with low productivity
None (0)	No agricultural land is affected	A project appears in circumstances of limited agricultural production/use of affected lands

Beneficial 1: Rural electrification
Magnitude: Installed capacity of the planned project
Importance: Current level of electrification in province

Very High (4)	Installed capacity ≥260 MW	Access to electricity ≤50 % of households
High (3)	Installed capacity 200 MW≤X<260 MW	Access to electricity 50 %<X≤70 %
Medium (2)	Installed capacity 90 MW≤X<200 MW	Access to electricity 70 %<X≤90 %
Low (1)	Installed capacity < 90 MW	Access to electricity 90 %<X<100 %
None (0)	No project	100 % of households have access to electricity

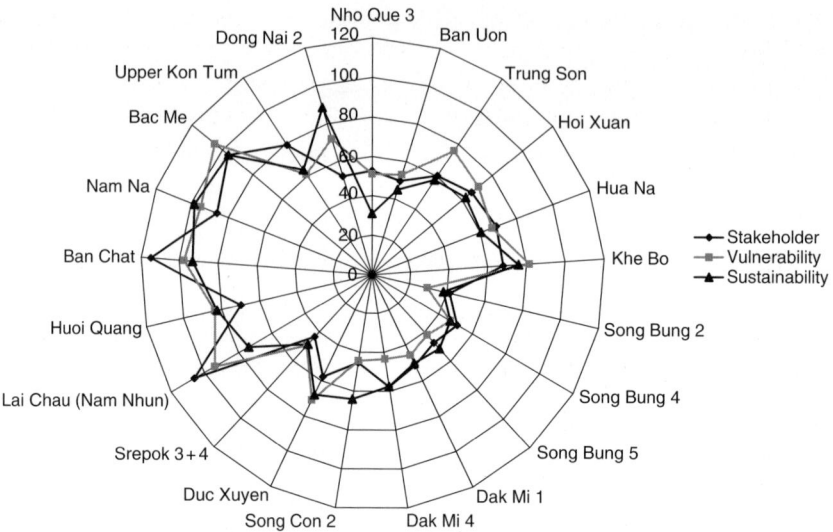

Fig. 1.1 Three assessments of potential social effects at selected sites, the NHP study

The approach can also be developed further than what was required for the NHP Study by ranking the potential projects solely from a social point of view. The methodology allows for easy comparison between projects, as Fig. 1.2 illustrates. The figure shows effectively how some projects can be high risk/high return, others low risk/low return (and intermediate, of course) in strictly social perspective. The curve "Alternative" in the figure stands for what can happen when beneficial parameters change through heavy investment.

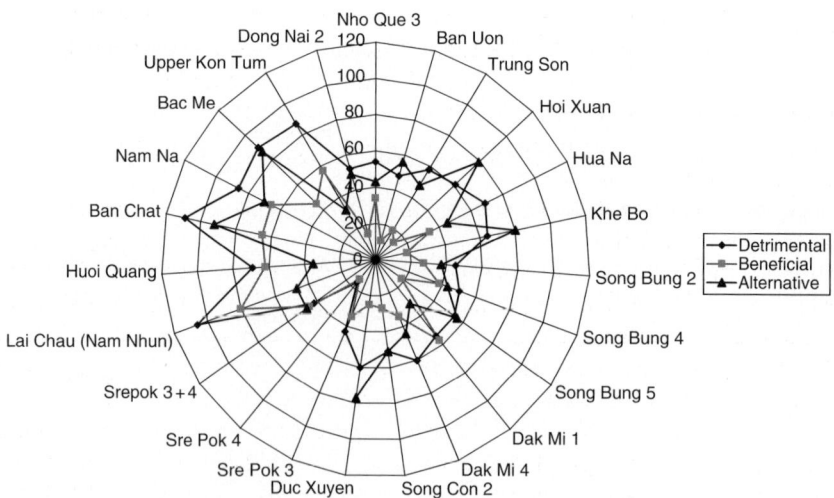

Fig. 1.2 Detrimental and beneficial scores for social issues, the NHP study

By looking at all potential projects in one river basin, the regional impact can also be immediately commented on in terms of high risk versus low risk from a social point of view. For an investor looking for minimal social destruction (and subsequently lowest social mitigation costs) such figures can provide good opening for further assessments.

1.7 The Hydropower Response to Sustainable Development

The attention to a sustainable prefix to development is emerging gradually to hydropower planning. This has been illustrated in the current chapter. Requirements have been specified both through the UNCED and WSSD, as well as through the WCD.

The WCD concludes values for an implementation strategy listed early in this chapter (equity, sustainability, efficiency, participatory decision-making, and accountability). The current study is confined to the relationship of hydropower development with sustainable development, showing how interconnected all these aspects are. The experience from the studied projects, to be detailed in the following chapters, is that the equity goal (poverty and gender) can be addressed through proper implementation, where a hydropower project becomes an asset for regional planning. The participatory decision-making is a necessary condition for this. It is also highly relevant for both efficiency and the sustainability of a hydropower project. And finally, it has direct bearing on the accountability called for by the WCD.

The examples raised in the current study cover projects over a decade of change in consequence of the global environmental conventions. They show a gradual adaptation to new conditions for hydropower planning; first through a request for better structured details on what was contextual information about a project's impact for the affected population, and then gradually a build-up in this demand to be of relevance for implementation. There is a more clear expectation that hydropower projects shall be multi-purpose. They should benefit also the Poor in a local population, and they should contribute to regional development in a country. To this comes social safeguarding through mitigation measures, nowadays more distinctly including resettlement and downstream situations.

With the increasing demand for social information, and action, a methodology development has followed. New techniques have been developed that are cost effective, probing into a mass of information on social situations rather than conventional impact assessments. The benefit is the ways in which these "quick and dirty" screening techniques have at least two effects; that stakeholders get involved and build their capacity, and that improved capacity is built to call for full-fledged social studies in times of need – when unexpected consequences appear.

The issue how to "find" the Poor in the social catchment of a project area has been addressed by looking at domestic water supply, drawing on the correlation between low level consumption and poverty. Involving local community stakeholders, not only the Poor, in the formation of social data has proven fruitful. Initially this was only as a technique to inform and open a dialogue channel, but then gradually it turned into a technique to mobilize and share decision-making on a number of specific issues, including those that are mitigation related. The stakeholder mobilization includes also other layers, not least regional within a country. Technically speaking it seems straight-forward to regard a large-scale hydropower project as a source of income and contribution to sustainable development. However, situations show huge variation depending on the international attention.

The next chapter will focus on data generation and information management under various circumstances. The purpose is to highlight how hydropower planning has come under the influence of the global environmental conventions through new demands and what efforts have been made/can be made to respond. Chap. 2 will reflect over the new methodologies applied in efforts to follow policy change, underline the needs for a new database, and how it can be dealt with.

Chapter 2
Social Catchments of Hydropower

A hydropower project can be associated with a social catchment impact area surrounding it. By social catchment is understood those areas where people relate directly to the river stretch concerned. The concept is used since the area identified can be very different from a local administrative unit; information management can be complicated since available data refer to an administrative unit, not to a social catchment.

New needs for data, information and participation in a sustainable development process appear as key issues in the present chapter. Experience is used from two different ways in the studied hydropower projects of identifying the social issues for those depending on river water.[1] The first section addresses the "old" conventional surveying approach; the second a process with involved stakeholders defining the key issues for their river water use. Illustrations come from several cases in Nicaragua, Costa Rica, Angola/Namibia and from Vietnam. Differences between the two approaches reflect the methodological development for data formation. It goes from survey with quantified data to controlled interviews and further to brainstorming workshops with consensus building. This process has taken place parallel with an increasing political acceptance to involve project affected people.

One of the illustrations of the latter kind comes from Northern Vietnam where stakeholders specified those project activities that were expected to generate positive effects on regional development. This is the most sophisticated process oriented example in the chapter. A notion of social catchment can be used in this case in order to underscore the fact that resource access for the participants is socially, not ecologically bound. It may concern production and marketing of farm products, or where graves and shrines are located.

The Nicaragua example in the current section represents the earliest one of the studies revisited. It began in 1993 at a time when consultancy firms fairly uncritically and on routine basis added a section on socio-economy in environmental pre-feasibility or feasibility studies. Very typically, the information given could be

---

[1] Experience from work for SwedPower with the La Sirena project is acknowledged. The contribution of the author was to lead the social studies within an EIA.

demography data derived from census reports, a bit of household economy including cash income, some production features, possibly supplemented with consumption patterns and a notion on ethnic (minority) situations. There are even examples of given information being partly disconnected from the study site. Targeted data gathering was unusual, and many reports written in the early 1990s today give a haphazard impression; whatever social information available seems to have been lifted into reports. Guesstimates in one report easily became truth in the next one.

The screening of key social information that was introduced to the Nicaraguan client was news at the time; the new thing being that there no longer was any effort made to deliver a "holistic" view but, in contrast, to highlight a few key issues. The first reaction from the client was disappointment with the short texts. Next reaction, however, was one of appreciation when the method proved suitable for comparing options and making priorities.

The second example of data formation without a stakeholder process mobilizing the directly affected people is the already mentioned Epupa feasibility study in Angola/Namibia, and circumstances where data formation could be done in a more profound way.[2] This was carried out by project staff. Stakeholders were mobilized at a few national workshops but regularly only in the project area, not regionally. It represents a major methodological development in that stakeholders become involved in data compilation and assessment.

The last example comes from northern Vietnam.[3] It goes one step further, into data formation, not only data compilation and assessment. The example is derived from a stakeholder driven process where potential sub-projects are identified by stakeholders. Like for the Angola/Namibia case this project is introduced in the previous chapter. Here an illustration is given of how a process of identifying needs and issues for compensation could be arranged.

2.1 On Data: Get What you Need Instead of Take What you Get

Two cases from Central America will illustrate how data scarcity can be met. First, the experience from Nicaragua shows how also scarcity of secondary data can still lead to a useful rapid screening. The observation is that data needs to be systematic in order to provide meaningful information. It is more constructive to have limited but targeted information than having unstructured bits and pieces that provide glimpses of social situations without coherent analysis. The second experience, from Costa Rica, concerns downstream effects of water regulation. The scenario is to minimize low flow, and how to identify the most vulnerable river stretches in terms of inhabitants' dependency on the river water.

---

[2] Norconsult was main consultant and Sida financed the feasibility study. The author's role was Team Leader for the social group; towards the end also for the EIA.

[3] Work in the Second Red River Basin Sector Project; Uplands. For ADB on RNE finance; see ADB 2005a.

2.1.1 Systematised Information in Case of Data Scarcity: The Case of Site Options in Nicaragua

The kind of information needed for pre-feasibility studies, and the rough formulation of an understanding of specific social issues for a hydropower project are discussed in this section. In most instances there would be a wealth of social information that is both unstructured and partial. In other cases, however, studies have had to fall back on the use of scanty data. In either case it is necessary to master an understanding of what kinds of data are needed. Structured search for information can allow for an identification of issues that reach beyond mere guesstimates within a small number of problem fields. The illustration below is drawn from a national inventory in Nicaragua,[4] where the project team met with little written information but had to produce pre-feasibility conclusions very quick.

The objective was to carry out a minimum of further detailed reviews and studies, and still arrive at a ranking of the alternative projects with respect to economic and environmental merits. A methodology was developed with a strategy for social data management. Only data on key issues was compiled. Data was structured according to themes (in later stages, 2000, to the study parameters), and systematically combined with environmental factors. Relevant engineering and economic factors were addressed parallel to the process of combining social and environmental issues. Taken together all four factors formed a ranking profile of a potential project in relation to alternative or supplementary sites. This approach was refined stepwise and applied in both assessment and training projects. The training took place in response to capacity building needs, as will be accounted for in Chap. 4. The current section gives the first step in such a work process; identifying what kind of data might be available, and how to generate supplementary needed data. New at the time was to move away from the practice of presenting whatever information might be available and instead restrict data collection to issues that had been specified in advance. This selection of issues was based on general experience from earlier projects.

The restricted data collection targeted four dimensions for evaluation: technical, economic, environmental and social. The project also developed a way to combine them. Combined Ranking between these four dimensions of potential hydropower projects formed a first approximation of a holistic understanding based on a small number of indicators. Rapid follow-ups were conducted through structured open-ended interviews in order to improve the basis for evaluation through performing the following:

- Further discussions with stakeholders and collection of additional information
- Further detailed review of previously collected and additional information
- Site visits to project sites and reservoir areas
- Cost estimates
- Development of economic indicators

---

[4] The prefeasibility Rio Viejo Project, Nicaragua (1993) and Hydroelectric Survey, Nicaragua (2000).

- Social and environmental assessments
- Technical/Economic and Environmental/Social screening
- Combined Technical/Economic and Environmental/Social ranking.

The analysis of available sources plus calibration according to the first two bullet points above proved to be a difficult step. This was due to the problematic physical access to the project locations. Four of the potential eight sites were actually not visited at all. Therefore the decision-making basis for pre-feasibility level was the systematic collection of data from available sources. Additional information could be collected in Managua, the capital city, without going to the field; in particular since aerial photos of high quality were available for all project sites. The access to the aerial photos had to compensate for the drawbacks in the site inspections. General hydrological data, but of uneven quantity and quality between the different rivers and projects, could also be purchased in Managua. However, missing site visits meant that no interaction with local stakeholders took place in half the potential project sites.

Eight potential sites had been selected for further detailed analysis in the project. Data was accordingly collected on each of them. One of them was Corriente Lira that will here illustrate data formation quality. The following is a summary from the author's own notes and contributions to the Hydroelectric Nicaragua report (2000), showing what kind of data was generated:

> For the example of Corriente Lira the study gave the following. It is a project site located on the Coco River, about 9 km upstream of the community of Wivilí with around 5,000 inhabitants, and about 2 km downstream of the confluence with the Cuá River. It was identified in the 1980 Master Plan study, in which a project with a high dam and a large reservoir was proposed. The project was eliminated, however, and was not included among the selected projects on the ranking list. A later study proposed a run-of-river plant with a much smaller dam at the same site, and it is on that basis that the coarse screening was performed.
>
> Available data included hydrology (12 years recording), geology (one master plan), environment (from aerial photos). Also the 1:50,000 topographical map with 20 m contour intervals was available but no other mapping. The social characteristics were estimated from maps (agro-forestry, coffee, subsistence agriculture, population density, number of inhabitants in the catchment, and physical infrastructure). Observations were made during site visits to one town and one village in the project area. This gave average household size, disease panorama, sanitation and education information for the project area, plus discussions on poverty and compensation through a couple of PRA meetings. The Corriente Lira site visit was to have an impression of the reservoir area, to visit one village that probably will be inundated in order to see its relations to the river, and to form an impression of the downstream situation at Wiwilí town.
>
> The area was heavily struck by the hurricane Mitch in the end of the 1990s. Most of the population left have returned. Before that it was a major war zone during the civil war 1980–1989, with troops from Cuba and the Dominican Republic deployed there. At that time most trees had been cleared, and populations settled in villages. Marañosa is one example, established in 1985.
>
> Agricultural production is formed partly by industrial coffee production (enterprises) and partly by livestock, maize and beans production. The village of Marañosa was visited and a PRA carried out.[5] This village was constructed in 1985 during the war. Inhabitants come

---

[5] PRA, participatory rural assessment, is a notion covering a mix of participatory assessment methods. They have in common that they are process oriented and look for stakeholder interaction in most cases. Each method's results should be cross checked through application of other methods.

from all over Nicaragua. Today it hosts 600 persons living in 65 houses. This means an average family size of around 9 persons. The figure is high compared with the national average
of 6 persons. As in entire Nicaragua there are more women than men (53 %). Average age
is said to be high in the village. The large size families are explained locally so that fairly
old people have become anxious to have many children. People feel poor, and say that they
can eat one full meal a day.
There is good quality drinking water available, taken from a borehole in Kilambe, 9 km
away. The river water is not used; it is seen as heavily contaminated. The nearest health
centre is in Malekong, 7 km away. Transport is good since Marañosa is located on the
Wiwilí-Jinotega road, serviced by 5 buses per day in each direction.
The two dominating diseases according to the discussion are Dengue fever and malaria.
There seems to be satisfaction with health service. The issue of water borne diseases is
known. Latrine quality is seen as poor; a pit latrine project is asked for. People know about
the heavy use of different pesticides on the coffee farms; this is good reason to avoid river
water.
Most inhabitants are farmers, also keeping cattle and chicken. Main production is maize and
beans. More development support in the agricultural sector has been a major issue in the
local election campaign that is ongoing. Other development is not prioritized for the village.
Inhabitants feel that they need better housing ("more houses"). A health centre is also called
for. There is no electricity and no telephone. But there is a primary school with 94 children
and classes 1–6. After that level there is the opportunity to go to secondary school in Wiwilí.
School attendance is in the magnitude 80 %, the remainder staying at home to work on the
farm, etc. School leavers are few, with about 90 % passing through class 6.

Similar information, but of shifting quality, was produced for in all eight options.
The choice of data gathered was made on the basis of a preconceived model based
on already selected key parameters. Therefore a transformation of existing data,
including that from the project's preparatory phase was made into a data sheet for
all base-line information. The aim of this step was to build a platform to carry out
a formalized scoring of social/environmental impacts. The scoring could be based
on quantifiable parameters, and also on the expert's assessment of data, in order
to derive social/environmental preference scores. An elaborated application of this
method was later developed in Vietnam, as further discussed in Sect. 2.2. The new
element in the Nicaragua study was that the methodology opened for a qualitative
social/environmental diagnosis for each project option, based on pre-set parameters
supplemented with "Warning Flags" comments from stakeholders.

The selection of key social data in the Nicaragua project was made in accordance
with what proved possible to access at the time. In order to follow the model design,
parameters could not be shifted. In cases where data was not available; this would
prevent the comparison between sites. The only way to uphold the option to compare
different sites was to totally exclude parameters where data was lacking in some
cases. This led to the selection of the limited number of key parameters presented
below. Both the qualitative (how significant the parameter is) and quantity sides of
each parameter were then considered in data assessments, once the platform had
been established.

The following social parameters were chosen for the selected projects:

- People affected by the project
- Area of land lost (total land take; jointly with the environmental assessment)
- Public health/disease

- Cultural resources
- Water quality (jointly with the environmental assessment).

For each parameter an Impact Score was determined through a pre-set criteria adapted to the span of variation among the projects. Impacts were assessed either by use of measurable entities or rated by use of selected numerical scales for potential impacts in each category. The scoring criteria were in brief been as follows.

2.1.1.1 Number of People Affected by the Project

The number of directly project-affected people ("direct PAP") was estimated in most cases based on map reading (topographic map 1:50'). The indirectly affected people ("indirect PAP") were all the remaining inhabitants of the area to be affected by the project in socio-economic or cultural terms. The rough estimation of the number of indirectly affected people was made on the basis of number of inhabitants in a 3 km wide strip along the river stretch of the catchment area.

2.1.1.2 Area of Land Lost (Total Land Take)

The physical area size of land lost had been estimated in earlier project documents. The question about land quality was also entered through commenting on the productive capacity of land taken by the project based on maps. Practically all inhabitants had access to some productive land. Farm sizes varied considerably but with a median in the magnitude of 5 *manzana* (3.5 ha) per household (with a median size of 6–7 persons). The land tenure picture was complex, with combinations of landowners, tenure holders and co-operatives. In social terms the local populations could still be seen as relating directly to the natural tracts potentially affected by a project through loss of productive land.

2.1.1.3 Public Health/Disease

The screening of Public Health and Disease situation focused on two major issues in following comments from field visits. One was the health situation relating to waterborne diseases, and the other was community health with roots in social diseases (primarily STDs). The incidence of socially related diseases, STDs, was connected to population size and proportions between host population and project employees.

2.1.1.4 Cultural Resources

There was a range of artifacts to consider: significant archaeological and palaeontological sites, graves and other religious places, etc. Other physical constructions, such as houses, schools and other infrastructure were not included in this category but correlated with population and already represented above. Information was attained from scattered reports and from one key (archaeology) informant.

2.1.1.5 Water Quality

The assessment included impacts related to potential water quality changes, which might affect downstream use of the water. This aspect referred to the environmental part of the evaluation. The combination with social assessment included: (1) domestic and industrial use of water; (2) irrigation, recession agriculture, cattle watering; (3) downstream ecosystems.

2.1.1.6 Length of River Affected (Flooding/Diversion)

This criterion fell into the environmental domain. It combined upstream and downstream effects. It was followed by other environmental/ecological parameters in the assessment.

With this approach to focus on a few key questions and concentrate the effort to achieving information on them, made it possible to assess a number of options in relative terms. The project could report a list of the options according to social preferences, along with the agreed aim of that project to merely rank alternative hydropower site options. By combining the social information with environmental, technical, and economic, the hydropower options could be grouped into "baskets" of low risk/high risk and low return/high return in energy production terms, combined with mitigation costs. Data was only sufficient to make such relative statements, and insufficient to give an assessment in absolute terms. The technique was then consequently further developed in other projects discussed below. In the current chapter the remaining issue is how decisive the first-level data formation is for shaping assessment models.

2.1.2 Replacing Data Scarcity with Indicators: The Case of Downstream Development in Costa Rica

The downstream effects of a potential hydropower dam on Rio Savegre in Costa Rica have been studied since 2004.[6] The approach in the social part has been to identify a number of sectors connected with the river downstream of the dam site, on the basis of the land use in each of them. The social catchment of each sector has been given by land use patterns, and not through a selected strip at a fix distance

---

[6] The study is being carried out by Hydroconsult AB. The author is through ENS Consult AB sub-contracted to assess the socio-economic effects.

from the river like in the Nicaragua case. The aim of the method has been to outline networks of economic and production activities; in some cases (depending on
eco-tourism and rafting) quite close to the river, in others (ranching and livestock
rearing) further away.

This is a technique that is more detailed than what was carried out in Nicaragua.
The indirectly affected areas downstream have traditionally (in hydropower studies)
been defined as a strip along the river (cf. Map 2.1) even with the same width independent of land and water use. In the Costa Rica study this static approach has
been replaced with a more specific attention to the different water users. The method
offers more flexibility to identify low flow levels that may vary over season rather
than being locked to a certain percentage of high-level water flow.

A key observation attained through the method is that a hydrological and a socio-
economic assessment generate different results, when searching for that minimum
flow that must be upheld to meet local demands. The differences in the current case
are small but significant for the directly affected stakeholders. The low-flow needs
for hydropower production of electricity in terms of volume of water per second
do not coincide with people's needs in terms of combining water availability with
socio-economic requirements. However, there might be room for variation rather

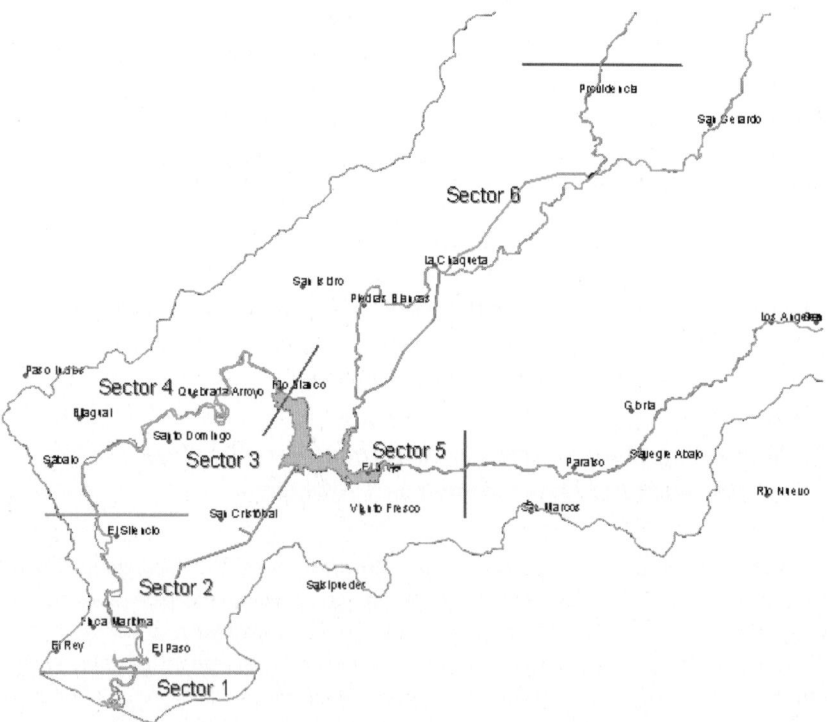

Map. 2.1 Illustration of how study sectors for the socio-economic assessment of key issues can be
made (map by Miguel Viquez Camacho).

than setting aside a constant figure for minimum flow. The results derived from this approach add up to transparency for the decisions to regulate low-flow levels. A variation from, say 10 % of total water flow of one or two percentages would mean huge benefits for electricity production. The detailed studies provide the answer if this is possible.

The river was divided into sectors, depending on ecology and resulting water use profiles through stakeholder consultations. Based on the joint experience of stakeholders and project staff several river sectors were identified based on dominating land use practices, including areas downstream of the reservoir as indicated on the map.

In the cooperation between local community stakeholders and project technical expertise (social and ecological) a number of parameters (initially 15, and then reduced to eight) were established to express all land use practices along the river. This process made use of the techniques developed in Vietnam. The parameters were also assessed for their significance by the local population. One set of assessments was made for each of the river sectors. Based on this assessment, the parameters were ascribed a score; High, Medium or Low, depending on the social importance of the river (Hydroconsult 2006). For instance, do fishermen in one sector have to rely on the river water in their own sector only, or can they fish elsewhere (upstream, downstream or in tributaries and streams)? The definition of scores for each parameter was as follows (Table 2.1).

A preliminary combined scoring per sector created a profile for each sector with regards to the eight types of land and water use; under Activity in the above table. This was achieved through multiplying the social and water scores. The resulting profile expressed the urgency in the demand for river water from a socio-economic point of view. The next step was to look into people's actual dependency on river water: are there alternatives, such as tributaries, streams, etc.? Few options mean high negative scores. High dependency levels are particularly sensitive in the case of future flow regulations. Finally, the achieved sensitivity in water availability was combined (by multiplication of the scores) into a combined figure for vulnerability. The following Fig. 2.1 shows the outcome for all sectors:

The most problematic issues for each sector can be illustrated in such a figure: The most crucial activity sectors could thus be located, so that the "hotspots" for low flow regulations from a social point of view specified. They consisted of the high score issues for each of the seven sectors (Sectors 3 and 4 on the map have been combined). The key issues identified in this way should be given prime attention when fixing low flow levels (in terms of volume; flow speed not addressed here). Mitigation measures should moreover be based on the findings.

There is a risk that minimum flows through regulation can be established based on other principles than sustainability from the social point of view. The minimum flow calculations are confined to existing consumption patterns. But the demand changes with new alternatives. For instance, the introduction of aqua-culture using river water is a promising pro-Poor proposal. It attracts women groups where members can find a supplementary income source. With the socio-economic technique poverty or gender perspectives add to the analysis by assessing the socio-economic

Table 2.1 Definitions of scores for social needs and water availability, Rio Savegre Costa Rica[7]

Activity	Issue	Range	Description
Fishing	Social needs	Low	Fishing is not important in the community
		Median	Fishing for recreation
		High	Fishing dominates in social life
	Water availability	1	There is too little water for any fishing
		2	Enough only for some fish species
		3	Fishing has to be done in this sector
Domestic	Social needs	Low	No use of river water
		Median	River water and streams available
		High	High dependency on river water
	Water availability	1	No river water demand
		2	Combining river, tributaries, sources
		3	Only river water available
Swimming	Social needs	Low	Festivals only
		Median	Weekend swimming
		High	Swimming more than weekly
	Water availability	1	No use of river water
		2	River water and streams available
		3	High dependency on river water
Livestock	Social needs	Low	Occasional use
		Median	Only dry season use
		High	All year use
	Water availability	1	No use of river water
		2	River water and streams available
		3	High dependency on river water
Rafting	Social needs	Low	No generation of local activities
		Median	Generating jobs/services
		High	Local rafting companies exist
	Water availability	1	No use of river water
		2	Seasonal use of river water
		3	Using river water year around
Aquacult.	Social needs	Low	No activity
		Median	External company active
		High	Local community group involved
	Water availability	1	No use of river water
		2	River water and streams available
		3	High dependency on river water
Irrigation	Social needs	Low	No use of river water
		Median	Occasional use
		High	Depending much on irrigation
	Water availability	1	No use of river water
		2	Combining river, tributaries, sources
		3	Depending much on irrigation
Tourism	Social needs	Low	Only by foreign persons
		Median	Creating local employment
		High	Possible use by local populations
	Water availability	1	No use of river water
		2	River water and streams available
		3	High dependency on river water

[7] These are preliminary scores done by the author to test the model; not the ones applied in the study referred to. They are presented here to illustrate an approach, not to account for results.

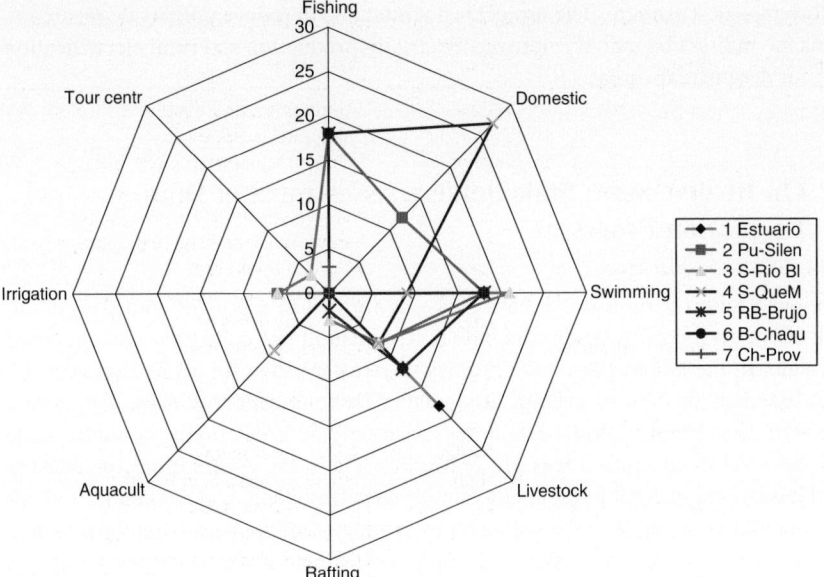

Fig. 2.1 Key socio-economic issues in low-flow circumstances

profiles for each river sector. Poverty reduction or gender equity projects can be located and "cost" in terms of minimum flow can be estimated.

The outlined approach can actually be generally applied, both for understanding downstream effects to be mitigated, and for prospective development towards multipurpose project effects. By focusing on the low-flow situations it becomes more transparent for the regional planner how to address downstream situations in a sustainable way. The approach allows for formulating the most critical periods in the regional development perspective. For a hydropower operation scheme valuable feedback is provided in the form of constraints and requirements for sustainable development. In the ideal case it will be possible for a hydropower operator to be flexible with regulation, and balance the optimal economic performance with the social and environmental demands, so that the sustainable development criteria are fulfilled. This is a more dynamic approach to low flow than merely setting aside a fix percentage.

This example from Costa Rica illustrates how downstream situations can be approached for regional development. It provides a certain degree of stakeholder participation – in data formation of course, and in defining the river stretches according to land and water use. Other than that the project carries out a rapid screening of water uses during low flow. The selected parameters are ascribed quality and quantity values, so that the critical stretches can be identified. The benefit is that both location and threatened river water use are identified. In this way a project can move quickly into feasibility and mitigation studies. With reference to sustainable development it concerns optimizing energy production with minimal social and economic disturbance downstream. The project is not intended to reduce poverty downstream. It has the indirect benefit of improved electricity production and rural electrification also for downstream areas.

2.2 On Involvement: Stakeholders Assessment of Data Formation Processes

Previous section demonstrated how data was formed in a project with the goal to rank alternative sites. A crude data base had to be built before any assessment could be made. In the Nicaraguan case, profiles were established for all hydropower options based on the final six criteria, along with a technique for how to ascribe scoring points to these criteria. With such an approach only internal ranking could be made and the level of resolution was pre-feasibility. The basic assumption was that the used indicators, in combination with similar ones from technical, economic and environmental viewpoints, were sufficient to group potential projects into promising, negative and in-between "baskets". The process worked in that the three groupings appeared in an orderly way.

The indicator use in connection with hydropower was improved as a build-up in the Epupa study in Angola and Namibia, already mentioned above and discussed further below. There is a basic difference. In the Epupa case stakeholders were involved in data formation (but not in study design). The survey methods and ways of representing social information in the indicator format followed the conventional approach, dealing with the social issues of stakeholders as passive objects. The methodology concern was now to be able to package and combine information to represent reality in a satisfactory way for needed decision-making processes in hydropower development and management.

Yet an alternative methodology approach, normally supplementary, is to empower the stakeholders and mobilize their competence into the assessment process. As will be demonstrated below, this approach can be viable in investment planning for small-scale projects.

As a general approach, stakeholder consultation has almost turned into a doctrine in development circles, and applied even when not suitable. Consequently a debate has been initiated that questions and holds back the approach (ADB 2004). Empowering stakeholders is applied in many ways, such as using the Participatory Poverty Assessment. Also in large-scale development projects the participatory approaches can prove effective at least in implementation phases. The NHP Study, introduced in Sect. 1.6 and discussed further below, incorporated several parallel stakeholder participation processes as an effort to get affected individuals and groups involved right from the beginning of a potential project, as mentioned in Chap. 1.

Learning from, and mobilizing, stakeholders, has been practiced in several of the currently described projects. The approaches in the Epupa project in Angola and Namibia, and the NHP Study in Vietnam are given in the following sections. These cases illustrate different phases of stakeholder involvement. It is not necessarily so that the methodology has been developed throughout the years 1995–2006, but the context in terms of regulations and political acceptance has changed during that time.

2.2.1 Stakeholder Participation in Data Formation and Evaluation of Selected Parameters (Angola and Namibia)

The local and regional studies of social consequences of the Epupa project on the border between Angola and Namibia took place in 1995–96 in the hot-air of international debate. At the time, stakeholder participation was confined to two aspects in a similar way to how the first stage of the NHP Study in Vietnam was conducted: Firstly, the end of each project phase was marked with stakeholder evaluations of results from commissioned studies. Secondly, data formation in these about one dozen special social studies involved participatory rural assessment, PRA, methods as much as possible. However, regular interaction proved impossible, not being allowed by authorities. As mentioned, the inception workshop on site was more or less carried through at gun-point. This authority performance, plus the initial presentation of the project implementation as pre-decided, sparked off an inflamed debate as mentioned in the previous chapter.

Based on the data needs specified in an earlier pre-feasibility study and also based on other experiences, a list of priority sub-studies was proposed to a stakeholder meeting at the onset of the project implementation. That list formed the basis for a one-day stakeholder interaction in Windhoek with open participation. The meeting was attended by about 200 interested persons, but notably without direct local community participation. The Himba community was instead represented by a legal advisor and by civil servants from the environmental ministry. The following list of priority studies was generated out of the workshop and carried out as commented (Table 2.2).

These studies were mostly carried out in close presence of the local community. But there was no continuous process involving stakeholders other than what emerged within the frame of the PRAs. Even though they generated good results, including studies of effects from an environmentally and socially friendly site selection, the political issue remained. Macro politics entered into the community assessment process in the form of ministerial statements and an initiative to carry out an alternative evaluation with the national university. The local community reaction was vicious. The tough political language from government was rejected through strengthened opposition among local leaders, and through a declaration that only the project's international staff was welcome into the potential project sites, but not the surveyors from the national university.

At the final stakeholder workshop, also with several hundred participants, the Himba community had sent two delegations, reflecting the polarization between developers and conservationists. The evaluation and interaction did not focus on project findings as much as on the national political process. Declarations were

Table 2.2 Stakeholder selected survey topics in Angola and Namibia

---

The social economy of Northern Kunene Region and the Provinces of Cunene and Namibe—a macro overview of potential impacts. Comment: This study gives the regional perspective for a potential hydro-power project both from Angolan and from Namibian point of view.

Resource management and pastoral production in the Epupa Project area (the Kunene drainage system from Swartbooisdrift to Otjinungwa) Part I. *Comment:* This study is on natural resource use and pastoral production systems in the surrounding areas of the potential project on the Namibian side.

Resource management and pastoral production in the Epupa Project area (the Kunene drainage system from Swartbooisdrift to Otjinungwa) Part II. *Comment:* This study is an elaboration of Part I, being more focused on the project area, both on the Angolan and the Namibian sides of River Cunene.

Report on field work performed in the Municipalities of Koroka and Tombwa in February, March and April 1996. The populations on the Cunene's right bank between Chitado and the river mouth, and the impact of the proposed dam. Comment: This study is a regional study of the communities concerned by the possible project in Angola.

Socio-cultural impacts on the Himba residents in the project area. Comment: This study focuses on the socio-cultural systems within the Himba as an ethnic group.

Report on first phase of archaeological impact evaluation of the Epupa Dam Project, Namibia. Comment: This is the first phase of the archaeological studies of the project area; it gives the first overview of the area and is confined only to the Namibian side.

Health and disease survey. Comment: This is the health survey conducted in Namibia.

Report on health impacts. Comment: This is the equivalent health study carried out in Angola.

Environmental impact assessment of dam site alternatives on the eco-tourism opportunities along the Cunene River in Namibia and Angola. Comment: This study covers different aspects of tourism industry.

Archaeological survey (Phase 2). *Comment:* This second survey is focused on the two main alternative sites Baynes and Epupa.

Preliminary report on a tour to the Angolan side of the proposed Cunene River hydro-electric dam. Comment: This study is on the attitudes of inhabitants in Iona Park, Angola.

Source: Norplan 1996

made by international environmental lobby groups opposing the project that the results were of no interest since the project to build hydropower dams was intrinsically bad and should be stopped on ethical grounds.

2.2.2 Stakeholder Demands for Regional Beneficial Effects (Vietnam)

Avoiding systematic involvement of local communities was also a feature at the beginning of the National Hydropower Plan Study, NHP, in Vietnam in 1999. However, engaging stakeholders was tried as the study got underway, and gained momentum as a constructive contribution to a planning process. Not only national but also regional workshops were held during its first stage, with stakeholder representation from selected authorities. The second stage of the NHP Study (2004–2006), in contrast, was in a position to allow considerable stakeholder participation in the form of differently targeted stakeholder workshops also locally. The Windhoek meetings in Namibia had been public gatherings with an urban elite, and the Vietnamese NHP Study first phase meetings had involved primarily civil servants. The participation in the second phase contrasted to both by following a careful stakeholder analysis, in accordance with the WCD advice. This means that representatives were sought for all principal stakeholder categories adapted to the Vietnamese situation. Table 2.2 demonstrates how the representation was set up. It shows the involvement at all the Study's stakeholder levels (with the exception of the national level which is oriented towards national administration). The original design is summarized in the following matrix (Table 2.3).

The categories underscored were represented at the consultations. The others could not be identified as stakeholder categories in the Vietnamese setting.

Stakeholder relations to hydropower project planning were different for the listed consultations. At national level project design and effects were discussed by key

Table 2.3 Stakeholder categories at the NHP Study workshops

Stakeholder Category (following UNCED and WCD)	1. National Workshops	2. River Basin Workshops	3. Province Workshops	4. Village Group Workshops
Women	Women's Union	Women's Union	Women's Union	Assessment participants
Children and youth	Nil	Youth organization	Youth organization	Direct participation
Indigenous people	Political representatives	Political representatives	Participation	Participation
NGOs	Selected NGOs	Provincial representatives	District level representatives	Commune representation
Local authorities	National line ministries	Province level representation	Province and District authorities and commune People's Committee	People's Committee
Employees, unions	Nil	Farmers' Association	Farmers' Association	Nil
Business	Represented through line ministries	Representation	Female SMEs	Nil
Technology, science	Professionals from EVN[8]	Nil	Nil	Nil
Land users	Nil	Farmers' Association	Farmers' Association	Direct participation

[8] Electricity of Vietnam, state-owned electricity company that also was the Project owner.

decision-makers and related sector interest owners. At second, river basin level, focus was on regional development and indirect effects of one or a cascade of hydropower projects. Stakeholders were recruited from province and district levels in affected river basins. The third series of consultations, Province Workshops, involved the districts and communes to be directly affected by any project implementation, with focus on mitigation. Similar consultations but with more emphasis on resettlement and downstream effects were carried out in a large number of Village Group workshops and consultations.

Thus issues for columns 3 and 4 in the table above were related to mitigation, while those for column 1 concerned policy implementation and project management. Column 2 issues were located in between, and concerned strategic development issues, notably poverty reduction. The issues and solutions that participants addressed reflect their social worldview; capacity to perceive problems and benefits, and ways to deal with them. The result may serve as an example of how participants see the social issues of a project. Of particular importance then is of course who the consulted stakeholders are: Their self image, background competence level and mandate ("constituency"). This is dealt with in Sect. 4.2 below.

The result of the consultations at river basin level shows how issues of key significance for hydropower project development are identified in a very close direct involvement of stakeholders in an early planning stage. Their anticipations for positive and negative impacts are given below. Table 2.4 shows the ranking of beneficial issues done by 92 participants through group work.

Each of the top four headings in the table above was structured into sub-issues. These give no other clear message than a general expectation that a large project brings social service, infrastructure and economic growth (including job opportunities and market availability). Apart from the generic list over development expectations, specific comments were given on regulating water source and water supply, development of eco-tourism, improved security and rural electrification, stabilizing underground water levels, improved irrigation agriculture and aquaculture, upgraded fishing, and specifications on what kind of infrastructure improvements can be anticipated (roads, school, health station, irrigation, clean water).

Table 2.4 Expected first-level beneficial consequences, the NHP study river basin workshops

Top beneficial issues	Percentage of total number of stakeholders
Economic development	86
Improved water quality	52
Access to more electricity	43
Infrastructure development	18
Environment improved	10
Local revenues	8
Education	3
Gender relations improved	2

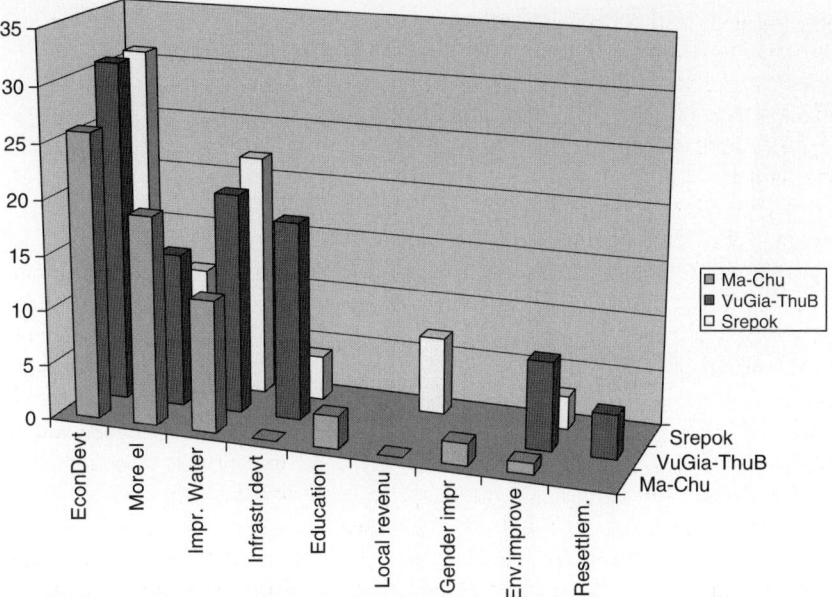

Fig. 2.2 The voting profile for anticipated beneficial consequences, the NHP study river basin workshops

Participants at the three workshops in Ma-Chu, Vu Gia-Thu Bon and Srepok River Basins voted for the different anticipated beneficial consequences. The results are shown in the Fig. 2.2 with the number of votes given on each expected beneficial effect from the planned projects in each river basin.

This priority list of anticipated development has been established by stakeholders with experience from regional planning and with a local viewpoint. Even if the output varies between the different river basins, a strong joint perception is that some of the water sub-sectors need priority, and that the multipurpose water use seems to be a logical expectation from a regional development point of view.

Depending on stakeholders' background their primary concerns varied: The importance different categories of participants (Farmers, Women, and Technical/economic staff) put on the 12 social issues (detrimental parameters, see Table 1.2 in Sect. 1.6 above) is given in the Fig. 2.3; it shows the percentage of stakeholders in each category that considered the respective issue to be of concern.

All categories showed primary concern over resettlement issues. Second ranked was the loss of agricultural land, farmer categories naturally being most expressive. Considerable concern was also expressed over unwanted effects of voluntary migration and over the plight of those partially or indirectly affected. Farmers were much concerned with potential poverty increase. Only technically or economically educated stakeholders gave attention to downstream effects. It should also be noted that ethnic and fishery issues were of no great concern, stakeholders seeing them as problems quite possible to mitigate.

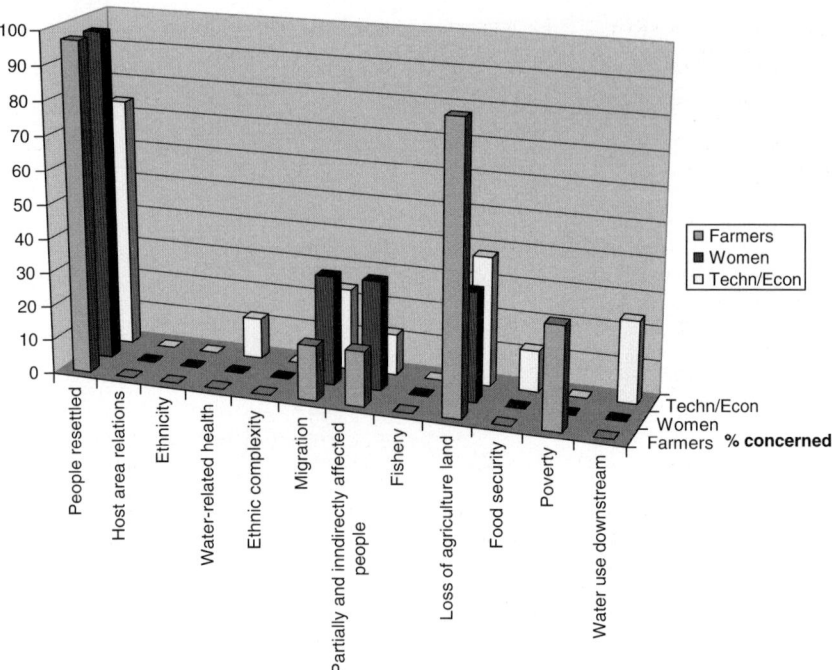

Fig. 2.3 Ranking of social parameters by stakeholder category, the NHP study river basin workshops

The output from the four workshops was not least, as also intended, a sensitization in affected districts and communes about the impact studies carried out. In particular the awareness about positive and negative consequences was raised through the ranking of perceived benefits and drawbacks. The ranking was subjective, for many participants based on first impressions. It should be seen as an outflow from the start-up of an interactive process for those localities where further planning must be continued. The workshop series was supplementary to other efforts where technical findings were presented and participants given the opportunity to carry out informed stakeholding.

Through the other workshop series, with stakeholders from the directly affected districts and communes, another 105 persons (of them 20 women) in leading positions were involved in a first round of consensus building about positive and negative effects for their respective communities of a possible implementation of a hydropower project. The evaluations all resulted in active involvement, with open and intense discussions along with professional information and facilitation. Participation and ranking of problems and benefits in the workshop series are summarized in the following Table 2.5.

A positive attitude to hydropower dominated strongly among the different consulted stakeholders. Participants were clear about potential problems and they were specific about identifying ways to remedy these. However, they were most keen to

Table 2.5 Participation and ranking of problems and benefits, the NHP study provincial workshops

Workshop location town (river basin)	Thanh Hoa (Ma-Chu)	Tam Ky (Vu Gia-Thu Bon)	Tuy Hoa (Ba)	Buon Ma Thuot (Srepok)
No. of persons (of these women)	18 (2)	28 (8)	24 (4)	35 (6)
Positive expectations (top number of votes)	Electricity for industry and business (13) Household electricity (11) Water for irrigated agriculture (9) Job creation (7)	Job creation (18) Improved life quality (13) Household electricity (11)	Job creation to enhance economy (18) Improved spiritual life (13) Meet demands for el in daily life (11)	Economic development (37) Irrigation and water supply for local people (20) Flood reduction (16) Improved el supply for local people (16)
Detrimental problems	Loss of real estate (16) Unstable life (15) Resettlement (10)	Loss of home and land (14) Resettlement (13) Shortage of productive land (7) Forest production (3)	Loss of property and part of forest used for honey production (14) Resettlement (10) Lack of land for cultivation (7)	Change of lifestyle (29) People must migrate (22) Loss of land for cultivation (21)

address the beneficial effects from an influx of capital into a poor region. Expectations on a positive trend brought about by the investment in a hydropower project were grouped around two arguments; that upgraded health and education services would follow as collective compensation for all those indirectly or partially affected, and that major capital flow would follow so that trade and job opportunities increase as spin-off effects.

These consultations in connection with the NHP Study represent a level where issues and solutions can be identified in interaction between technical staff and concerned people. But the consultation was not repeated so that it could allow stakeholders to build their understanding of the project. According to the client this was to be done during later project formulations. In the next section an illustration is given from outside hydropower but within IWRM, of how such an interaction process can be built; leading to informed stakeholder consensus based on both local stakeholders' and technical staff's expertise about local circumstances and technical consequences of investment and construction.

2.3 On Process: Stakeholders Directing Data Formation

The examples so far have illustrated an increasing involvement of stakeholder representation in a hydropower planning process. The original proposal for the NHP

Study to have open recruitment of stakeholders, going through an election process, was accepted only in principle but never implemented due to a client decision. This would be a next step towards upgrading the investment aspect into seeing a hydropower project as being multi-purpose, also beyond water resource management into issues like regional development and poverty reduction. The current section follows up on this latter issue; regional development and the role of a hydropower project. It links with the discussion from Chap. 1 of sustainable development and how development planning of such a large-scale project can pay attention to poverty reduction along with the regular criteria for sustainable development; that is to balance social, ecological and economic issues within a frame of what a project economy and technical capacity can permit.

The involvement of stakeholders in the NHP Study was restricted to consultations. There is one more step to be taken in order to live up the environmental convention work; to also involve stakeholders in a decision-making process. This section now goes into such a process application for mitigation with maximum poverty reduction. It gives the concrete example from Chap. 1 of the application of an IWRM where stakeholders made priorities based on their own competence as well as on outside expertise inputs. In this respect the methodology falls in the category of PPA, Participatory Poverty Assessment, as mentioned in Chap. 1. The experiences from water related project identification are taken up in the current study due to their relevance also for hydropower development.

The methodology developed in the Second Red River Basin Sector Project, Part A, Component 3, mentioned in Sect. 1.4.1, was to use IWRM as an instrument to maximize poverty reduction through an identification of the best suited water subprojects. It was designed to be useful in assessing potential mitigation measures as well, if need be. The principle was to combine what is possible with what is desirable: What is possible to implement from a technical point of view, based on external factors such as water availability, infrastructure arrangements and production cost of the mitigation measures, is first used to exclude a group of non-feasible (in technical respect) alternatives. Next step is to identify what might be or clearly is technically feasible. The environmental aspects of remaining proposals can then be assessed from environmental and from social points of view. The technique for the social considerations is illustrated below. The whole step-by-step process of identifying a group of preferred mitigation projects is followed and assessed by different stakeholder categories consisting of local authorities (sector interests) and of affected people. Decision-making over priorities and exclusions/inclusions is made on the basis of all information given by those directly affected.

The following example from the study illustrates this methodology to mobilize stakeholders into decision-making. It comes from two communes in upland Northern Vietnam. The task was to identify potential sub-projects within an IWRM framework jointly between villagers, commune administration, technical experts, and facilitation consultants. The output, proposed water sub-projects that were designed for achieving maximum poverty reduction, was compared with district and province level plans for rural development. This was, after the step-by-step identification process mentioned in Chap. 1, done at village level in 17 out of in total about 40 involved villages in two stakeholder selected communes; Bac Thong and Vo Nhai.

The task to zoom in on sustainable sub-projects from the general mandate to relate poverty reduction and IWRM involved five levels of technical expert assessments (water availability, technical feasibility, economic soundness, poverty effects and environmental consequences) at sub-project levels. The project set up a process with reporting these technical assessments to village level stakeholders. In this way a process of dialogue through SWOT analyses was established. It ended in a priority list. For the specific hydropower water sector the lesson is a technique to draw on local knowledge for project design.

2.3.1 Two Vietnamese Communes in Search for Potential Development Projects[9]

The outlined stakeholder process can be applied stepwise in order for three views to meet; those of local stakeholders, technical experts and provincial and district development planners. The case from Northern Vietnam concerns investment planning in water sub-sectors. Project planning should be done in such a way that the poor benefit as much as possible in this case. Hydropower planning is different, but there are lessons for project design. These do not concern project identification but identification of mitigation measures. The lessons can be supportive to efforts in hydropower where investment is planned to build sustainable regional development.

After the political decision to follow a procedure leading to the identification of a few sites for large-scale investment (as an alternative to distributing limited resources thinly over a region), the selection of sub-projects in the IWRM case followed three steps. *The first* comprised of consultations at province level with the ministry of agriculture to assess priorities already made with its province level structures. This interaction has led to the specification of one catchment per province to be targeted for investment. All catchments, of a size similar to one or two districts affected, were identified, assessed and ranked for priority by district level administrations and technical expertise. Given that data was formed on administrative, and not ecological, basis the output was a selection of two sites per district. *The second* step, described in further detail below, was a stakeholder driven selection of priority sites for each of these. It in turn generated two top-ranked sites, and stakeholders at local levels (villages and communes) built consensus about priorities. This was done in two steps: (1) on the basis of the stakeholders' current knowledge, and (2) after detailed consultations with the project's technical experts. This consultation included technical assessments in five dimensions of all identified sites (water availability, technical feasibility, economic feasibility, environmental soundness,

_____

[9] This section draws on a study carried out within the Second Red River Basin Sector Project, Part A, Water Resources Management, an ADB project financed with a grant from the Royal Netherlands Embassy. The author acted as Component Leader for Component 3: Investment Planning in the Uplands; see Hjort-af-Ornas and Bich Ngoc (2004).

and poverty reduction potential). It is described in further detail below. *The third*
step was the stakeholder interaction between both national/provincial administrations (from step 1) and local (village, commune, district) representations (from step
2) in order to build consensus around top ranked options, following the same five
evaluation criteria.

The process of selecting top-ranked potential sub-projects for investment planning accordingly followed three steps in order to safe-guard both the utilization of
expertise, local knowledge included, and the accountability to all stakeholders in
proposed projects.

**Step 1. Stakeholder involvement in the selection of priority catchments at
province level**
At the onset, two provinces had been selected by the client; Bac Kan and Thai
Nguyen. The first step in the stakeholder selection process was consultation and
consensus building about priority catchments in these provinces. All catchments of
a magnitude 100 km^2 were identified by the water resource experts, and basic socio-
economic information was compiled covering poverty levels, demography, production performance and productivity yields within the catchments. This data formation had to be estimates with the help of relevant province and districts since all
statistics are kept for administrative units without following ecological boundaries.
Province and district levels stakeholders from relevant sector ministries, political
decision-making bodies, Women's Union and Farmers' Association interacted and
built consensus about one priority catchment in each province.

**Step 2a. Stakeholder involvement in the selection of priority catchments at
district and commune levels**
A similar process then followed, but now aiming for inventories of all small catchments within each of the selected two. At this point the inventory covered all small
catchments of the magnitude 10 km^2, bringing the size down to be on par with commune level. Stakeholders from all identified districts and communes participated in
a consultation meeting, and built consensus again; now ranking these catchments.
Two small catchments were selected for each of the two identified districts. For each
of these catchments a full inventory of all potential water sub-projects followed.

Step 2b. Village consultations
Still within Step 2, village consultations were carried out in all about 40 villages
that were situated in the remaining areas for potential sub-projects. Twenty five of
these, the most significantly located ones, were consulted in an intensive process,
including training of trainers who carried out 3-day workshops in each village in
order to build consensus about village level priorities among the basket of potential
sub-projects for further planning.

The technical assessment of the remaining potential sub-projects was made in a
systematic way by recruited international experts as described below before being
reported to commune and district administrations and stakeholders. The five sets of
assessments were each summarized into a score for each optional site as follows.

Step 3. Linking stakeholder process results into district socio-economic plans
The comprehensive stakeholder interaction process was not only intended to deliver data on potential sub-projects. It also created a dialogue that was fed into regular administration planning for sustainable development. This was done through yet another stakeholder interaction process between three partners; the local communities that in the end had received a priority mandate, technical experts, and local government. Through a facilitated interaction workshop recommendations from these three perspectives were merged towards one joint set of priorities. This process was so successful that the government decided to finance the finally proposed sub-projects over regular budget (not with donor support).

The technical assessments that were reported on to village, commune and district stakeholders in different phases of the project, are directly applicable to hydropower planning. All hydropower projects should be multi-purpose, and the supplementary purposes can be assessed in the just outlined way. The five dimensions, water availability, environment, economy, technical, and social/poverty, were treated as indicators. Water availability and economy had a cut-off capacity, disqualifying non-sustainable proposals. In the end the five indicators were combined, they could be given different weights, and decision-makers received a transparent information base.

The following summary of the methodology that was developed in the technical assessment parallel to the just mention process, and reported to the stakeholders there. At a concluding workshop the two findings were compared, a SWOT based ranking was made, and consensus was reached about priorities. These priorities were incidentally not of interest for ADB for investment. But local government appreciated the process and its result. They therefore managed to finance top-ranked proposals.

There are five independent assessments made.

- Water availability is of course conditional. An established model, often used at the Ministry of Agriculture and Rural Development was used. It gives conservative results, so that there is no risk that a cleared proposal on water availability grounds will fail in this respect
- Environmental impacts are estimated and ascribed scores depending on how serious negative effects are. Natural reserves and national parks are forbidding elements in the analysis, should they appear
- Economic features were represented by a calculated BCR for each potential sub-project. Negative options were blocked in the discussion. There was much concern among stakeholders that this approach was misleading and not reliable
- Technical assessment was difficult to make. The result was generally accepted by the stakeholders as being realistic. Options claimed to be non-feasible in technical terms were lifted out from the list of priorities
- Social/poverty assessments of each project were made in terms of negative potential impact.

These five technical considerations were weighed together, similarly to the NHP Study developed technique. A ranked short-list of potential projects was created.

This could be discussed by stakeholders with first-hand information, so that good feedback was provided instantly.

2.3.1.1 Water Availability

The calculations of the scores for water availability followed the assessment based on a standard model for crude annual averages. This was considered a safe and conservative approach that would guarantee that passing options definitely were sustainable in the water availability dimension. The most reliable situations were ascribed the highest scores. The scoring scale was designed as follows:

- Difference between availability and requirement (in m^3/sec) >3 means score 3
- Difference from about 1 to 3 leads to score 2
- Difference distinctly below 1 gives score 1
- Difference below 0.1 is score 0.

Summarized into a table of potential sub-projects this gave the following results (Table 2.6):

This table is from the 2RRBSP, Part 2, Component 3 preliminary final report (ADB 2005b). Its purpose here is only to illustrate what the outcome from the hydraulic model looks like (net water availability), how scores are ascribed on the basis of very reliable water access, barely enough water, and availability in between. In diagrammatic form the result is as follows (Fig. 2.4).

The study shows that water shortage was no problem for all potential sub-projects. The availability varied.

2.3.1.2 Environment

The scale was designed so that a high score was beneficial in the assessment of a sub-project. Three factors were considered:

Table 2.6 Water availability in 11 potential IWRM sub-projects

Sub-project Number	Type of structure	W avail. Minus w requirements	Score	Calibrated score (S × 2)
1 Quan Che	Dam	3.95	3	6
2ab Sui Bun, Trang Xa	Weir/PS	0.94	2	4
3 Dong Ra	Weir	0.03	0	0
4 Xom Pho	PS	3.89	3	6
5Binh Long2	Weir	0.06	0	0
6 Vai Tai	Dam	6.31	3	6
7Na Phai	Weir	0.01	0	0
8 Va Vat	Weir	0.19	1	2
9 Po Deng and V. Ngan	Weir	0.04	0	0
10 Na Bo and Khuoi Qua	Weir	0.04	0	0
11 Na Lat	Rev.	0	0	0

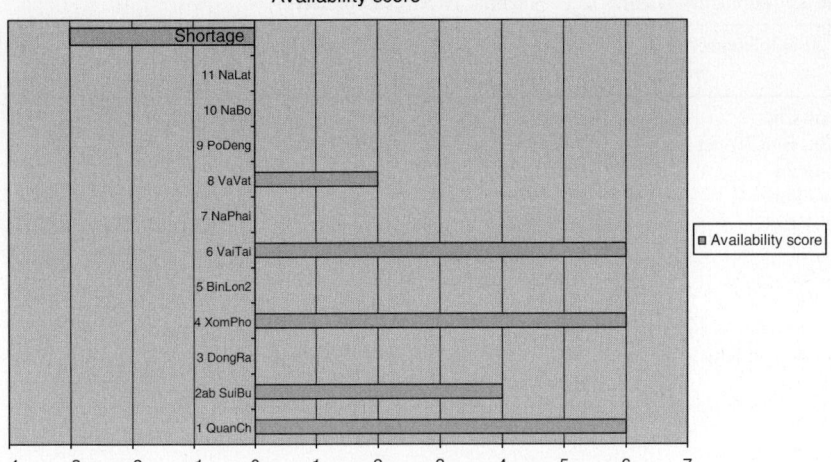

Fig. 2.4 Successful sub-projects from water availability point of view

- Physical
- Ecology
- Social

Each potential sub-project was assessed according to these parameters. Low impact was considered positive for the proposed sub-projects assessed. The result was summarized into:

- No negative impact (score 3)
- Little impact (score 2)
- Moderate impact (score 1)
- High impact (score 0).

The total score from the environmental assessment of potential sub-projects was made up of the sum of the scores for each potential sub-project. The scoring result with the combined score is summarized in Table 2.7 and Fig. 2.5:

No potential sub-project had severe negative environmental consequences; The profile is given in the graph; high scores = lowest negative effects.

2.3.1.3 Economic

The economic assessment of the potential sub-projects was brought to calculations of Benefit/Cost Ratios. These BCR were used for scoring the potential sub-projects as follows:

- BCR much above 1.2 means score 3
- BCR around 1.2 gives score 2

Table 2.7 Ecological features of 11 potential IWRM sub-projects

Sub-project Number	Type of structure	Physical	Ecological	Social	Sum
1 Quan Che	Dam	3	2	2	7
2ab Sui Bun, Trang Xa	Weir/PS	3	2	2	7
3 Dong Ra	Weir	2	2	1	5
4 Xom Pho	PS	3	2	3	8
5Binh Long2	Weir	3	2	3	8
6 Vai Tai	Dam	3	2	1	6
7Na Phai	Weir	2	2	3	7
8 Va Vat	Weir	2	2	3	7
9 Po Deng and V. Ngan	Weir	2	2	3	7
10 Na Bo and Khuoi Qua	Weir	3	2	2	7
11 Na Lat	Rev.	3	2	3	8

- BCR around 1.0 makes score 1
- BCR less than 1 is 0.

Summarizing the results into a table gave the following results (Table 2.8 and Fig. 2.6):

For groups of sub-projects under each heading the highest BCR is listed in the table. The weighted score means doubled value in order to calibrate the magnitudes with the other assessment scores.

From a strict minimal viewpoint the economic profile, as expressed through BCR, is given in the figure. The more detailed assessment shows that it is 2a and 5b providing the highest scores. The combined "package" of sustainable sub-projects was formed by:

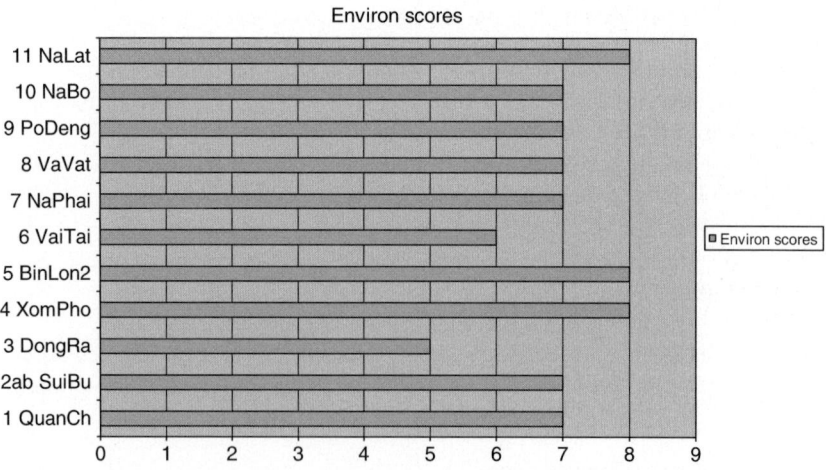

Fig. 2.5 Successful sub-projects from ecological points of view

Table 2.8 Economic features of 11 potential IWRM sub-projects

Sub-projectNumber	Type of structure	Benefit Cost Ratio	Score	Weighted Score (x2)
1 Quan Che	Dam	0.63	0	0
2ab Sui Bun, Trang Xa	Weir/PS	2.90	3	6
3 Dong Ra	Weir	0.95	1	2
4 Xom Pho	PS	0.26	0	0
5 Binh Long2	Weir	2.37	3	6
6 Vai Tai	Dam	0.36	0	0
7Na Phai	Weir	0.68	0	0
8 Va Vat	Weir	0.27	0	0
9 Po Deng and V. Ngan	Weir	0.28	0	0
10 Na Bo and Khuoi Qua	Weir	0.21	0	0
11 Na Lat	Rev.	–	0	0

1. Upgrading Quan Che reservoir and Canal system (Dan Tien, Binh Long communes)
2. Upgrading Suoi Bun weir and Canal System (Trang Xa commune)
3. Group of irrigation systems in Dong Ra (Dan Tien commune)
4. New Pumping station Xom Ven -Na Soc (Binh Long commune)
5. Group of irrigation systems in Binh Long 2 (Binh Long commune)
6. New investment of irrigation system Vai Tai (Phu Thuong commune).

Details of the sub-sub-projects are given in the section below combining the scores.

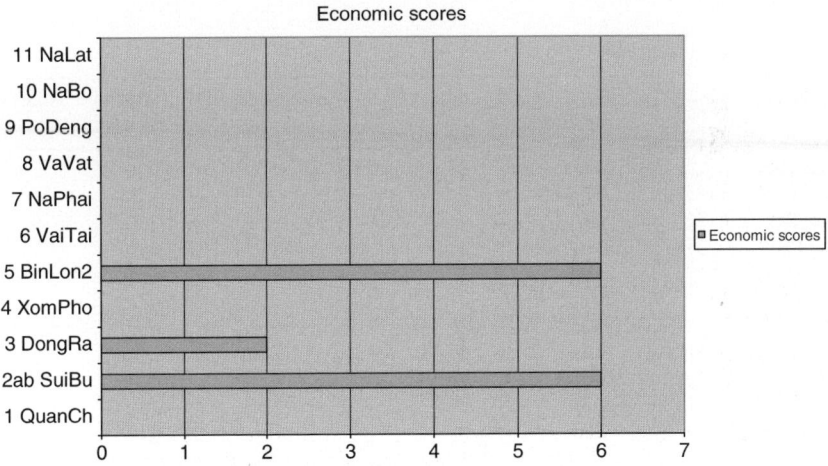

Fig. 2.6 Successful sub-projects from an economic point of view

2.3.1.4 Technical

In the assessment of findings from the technical evaluations of potential sub-projects two factors were considered:

- Size of area/increase in productivity (Scale)
- Degree of impact (Improvement).

The technological assessment was more complicated than water availability and environmental, since it included hypothetical constructions. Therefore, a technique of combining quality and quantity scoring was applied. In some instances the effect would be increased productivity and not expanded irrigated agriculture land. This was then considered instead; Improvement became a quality parameter to be combined with the scale of a project. The ordinal scoring was arranged so that large area or significantly increased productivity gave the highest score. In the same way the assessment of how significant the potential sub-project was addressed. The greatest change in the current situation gave the highest score. The two scores were multiplied in order to give the combined score (Table 2.9 and Fig. 2.7):

All potential sub-projects were technically feasible. The scoring was difficult to do for the technical profiles. The design was to show positive effects in the form of magnitude (for instance area gained) and importance (how significant a contribution the sub-project would have for existing systems). Their differences in magnitude and importance for the local communities involved gave rise to the technical score profile above.

As with the economic assessment the technical one went into further detail. This is specified in the last section where the different scores are combined.

Table 2.9 Technical features of 11 potential IWRM sub-projects

Sub-project Number	Type of structure	Scale (A)	Improvement (B)	Weighted Score (A×B)
1 Quan Che	Dam	3	3	9
2ab Sui Bun, Trang Xa	Weir/PS	2	2	4
3 Dong Ra	Weir	1	3	3
4 Xom Pho	PS	1	1	1
5 Binh Long2	Weir	2	1	2
6 Vai Tai	Dam	0	0	0
7Na Phai	Weir	2	3	6
8 Va Vat	Weir	2	3	6
9 Po Deng and V. Ngan	Weir	2	3	6
10 Na Bo and Khuoi Qua	Weir	1	2	2
11 Na Lat	Rev.	0	0	0

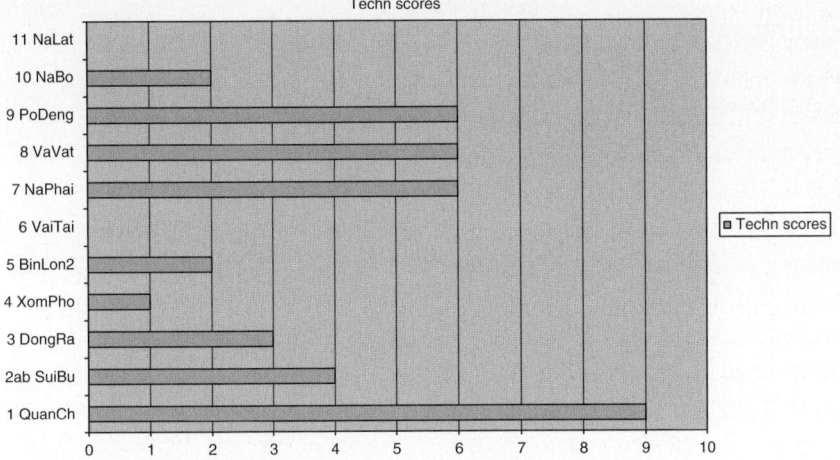

Fig. 2.7 Successful sub-projects from a technical point of view

2.3.1.5 Social/poverty

Four factors were used to indicate the level of poverty/vulnerability:

- % using dug wells ($< 70\% = 3$, $70 - 80\% = 2$, $80 - 90\% = 1$, $> 90\% = 0$)
- % poverty ($< 10\% = 0$, $10 - 15\% = 1$, $15 - 20\% = 2$, $> 20\% = 3$)
- % using dry pit latrines ($< 50\% = 3$, $50 - 60\% = 2$, $60 - 80\% = 1$, $> 80\% = 0$)
- % Kinh village inhabitants ($< 10\% = 3$, $10 - 20\% = 2$, $20 - 30\% = 1$, $> 30\% = 0$).

The four factors were added in order to achieve a combined social score (Table 2.10 and Fig. 2.8):

Table 2.10 Social features (including poverty) of 11 potential IWRM sub-projects

Sub-project Number	Type of structure	Poverty	Dug wells	Pit latrines	Ethnic minority	Combined social score
1 Quan Che	Dam	2	2	2	0	6
2ab Sui Bun, Trang Xa	Weir/PS	2	3	1	2	8
3 Dong Ra	Weir	3	3	3	2	11
4 Xom Pho	PS	1	2	1	0	4
5 Binh Long2	Weir	1	3	2	1	7
6 Vai Tai	Dam	0	0	0	0	0
7Na Phai	Weir	2	0	0	2	4
8 Va Vat	Weir	2	0	0	2	4
9 Po Deng and V. Ngan	Weir	3	0	0	2	5
10 Na Bo and Khuoi Qua	Weir	1	2	0	2	5
11 Na Lat	Rev.	1	2	0	2	5

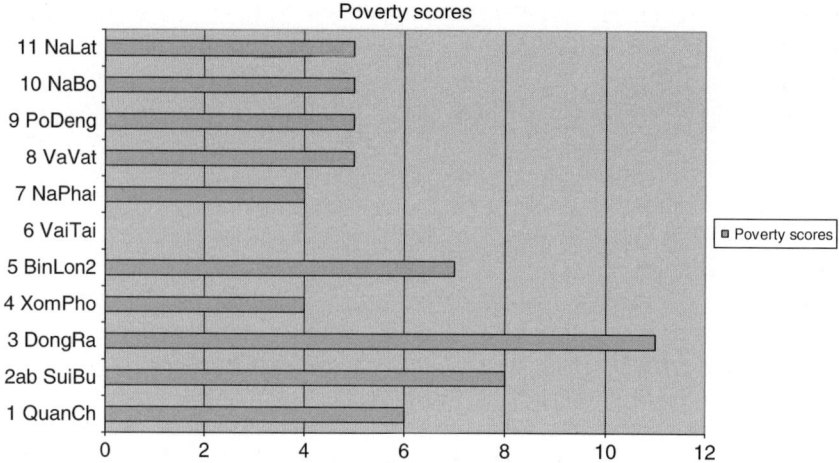

Fig. 2.8 Successful sub-projects from social points of view

2.3.1.6 The Combined Score

In this way the stakeholders ascribed each potential sub-project a profile based on the five assessment dimensions; water availability, environment, economy, technical and poverty. The following graph gives this plus the combined score for each of the potential sub-projects (Fig. 2.9). (Combining the five scores for each potential subproject can be done for instance by adding the scores for each potential sub-project. This is done in the next graph.)

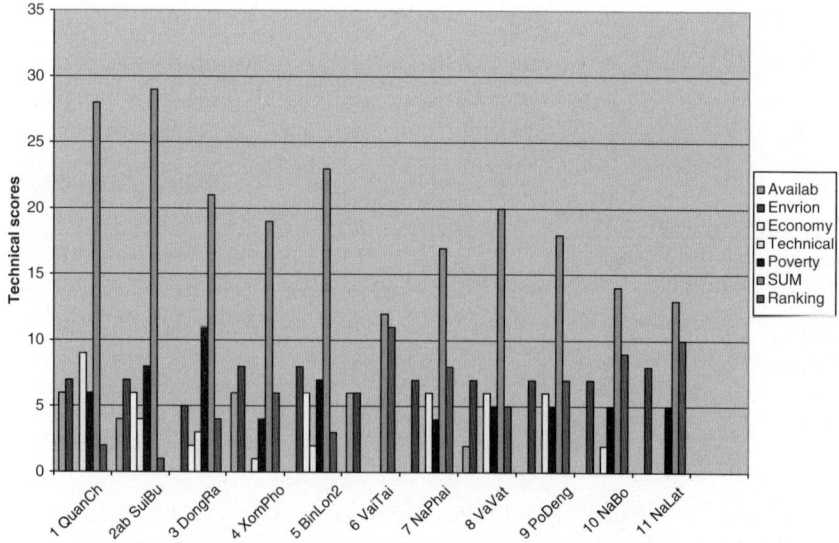

Fig. 2.9 Combined positive score for 11 potential IWRM sub-projects

The result illustrates that the potential sub-projects could be: 3. Dong Ra; 2ab. Suoi Bun; and 5. Binh Long 2. These were the most attractive ones from both economic and poverty points of view. From the view of all five assessments combined the top three were 2a Buoi Bun, 1. Quan Chi and 5 Binh Long 2. The result became a platform for decision-makers to discuss various considerations. The design also allowed for ascribing different weights to the five assessments (as well as, of course, a cut-off such as not allowing non-economic options).

The described stepwise approach opened for an opportunity for stakeholders to be involved in decision-making by accessing technical assessments at a point in time before sites had been decided. A similar approach could be used within a hydropower project; for detailed site selection, for location of infrastructure in both construction and operation phases, and not least mitigation like rural electrification, education and health facilities access. The WCD recommendations, as well as most segments of the international community, emphasize that mitigation should not be done through cash payment, and should be long-term and framed into sustainable development. Planners and developers often consider these requirements unrealistic. The thinking behind the example just given should counter that. It shows a transparent and low-budget way of identifying a basket of options, which, however, takes some time. But the reward is a transparent process, and one that involves all stakeholders to reach a consensus priority. Investment planning can be included, and thus a feedback into the hydropower project budget post for mitigation is achieved.

2.4 Social Catchment, an Issue for Stakeholder Involvement in Planning Process

In the examples given above in this chapter, the affected population of a hydropower/water management project has appeared with varying roles. The actual definition of who is affected is a major concern over responsibility for hydropower projects. It can have far-reaching economic implications in the form of mitigation measures. A hydropower project is a focal point for policy formation at both national and international levels. The issue from Chap. 1 comes back; it is so easy to drift from principles for sustainable development into financial and cost-alone principles for assessing a project's potential. Careful stakeholder participation opens both for assessment of data and information compiled for an analysis, and furthermore involvement in the analysis itself as in the last IWRM example from Vietnam. The mobilization implies involvement in the decision-making as well, which again is an activity with budget implications.

The examples varying degrees of involvement. The livelihood conditions and dependency on the river water vary. In the study in Nicaragua such dependency on river water was outlined: The three key sub-sectors proved to be agriculture, fishing and domestic water supply. Seasonal dependency on river water for these activities was described, including the specification of migratory routes for livestock watering. In the Angola/Namibia study this latter issue was all dominating and more complicated. Due to the dry circumstances the years vary, and the project assumed a worst-case scenario. Also here migratory routes were mapped, covering a couple of decades. Another water sub-sector was added: The role of the river as a line oasis providing emergency shelter (shade), food (dates) and water. The river dependency in this case included most of the Himba ethnic sub-groups relating to the studied stretch of the river. The resource base was broader than the river catchment. In other instances it might not be far-way pastures but mines and towns, sources for remittances and other resources flows. Such areas need to be included and form a social catchment. In the case of the many optional sites in the Vietnamese study, the social catchment was defined from infrastructure, markets and social/trade interaction. A rough statistical outline was made, based on the ecological boundaries of a river basin rather than on the administrative ones of communes and districts.

All approaches lead to a major methodological difficulty. Socio-economic data is of course organized on the basis of administrative boundaries, and these do not follow ecological ones. For all projects this has meant a negotiation of statistical information, with approximations for instance of a social catchment area with whole or parts of communes. The topographic map has in all cases been a prime tool to locate the social catchments and thereby to identify the prime target groups for the social assessments.

One difference between the three country situations has been degrees of verification with stakeholders. In Nicaragua there was no verification process at all within the project, while in Namibia the verification was made by urban, mostly white, liberals. In the case of the National Hydropower Plan Study in Vietnam, civil servants and other representatives of the political structures made most of the assessments. It is hard to see alternative approaches, but the issue remains a basic one. Therefore another Vietnamese example was brought up from small-scale IWRM investment planning, where stakeholders not only were consulted but also significantly influenced decisions towards maximum poverty reduction impact. The social catchment in particular identifies the directly affected and those who are indirectly affected in one way or another. This is normally a large population, and many national policies (the Vietnamese included) shun away from recognizing them as a stakeholder category positioned to demand compensation.

An experience from the many projects studied is, however, that this is a key concern for a sustainable approach to energy production in the sense agreed in the WSSD. The issue is then not safeguarding against negative effects as much as drawing on the huge capital flow into a region. The economics of a hydropower project take on a broader profile; the concern should be not only with BCR for the project, for instance, but also with regional investments in environment and social issues, and notably in poverty reduction.

This concern with regional development is addressed in the next chapter. It deals with one of the most logical forms of mitigation for hydropower projects. This is providing stakeholders a share of the electricity that is the outcome of a project. Chap. 4 then touches on one central hydropower consequence, the plight of resettled populations. The issue of stakeholder participation is returned to through the plain question: *Who is actually speaking up for the various stakeholder categories?*

Chapter 3
Hydropower and Regional Development for Poverty Reduction

The social catchment dimension of rural development with a hydropower project has focused on data and information with reference to how people relate to the river. It stressed the importance for stakeholders to involve in order for their region to benefit, if possible. This chapter goes further into the risks and benefits. On the one hand there are the increased regional risks with hydropower projects, and on the other their potential to make indirect contributions to regional poverty reduction. There is both an expectation and an assumption that the design shall meet poverty reduction demands through including a set of rural development measures. These fall notably within the economic activities irrigation, fishing, transport, and flood control, so that a hydropower scheme becomes multi-purpose. The combined effects have become a compulsory target for hydropower planning today. The world-wide attention has significantly contributed to these requests. Contributing are probably the global conventions, such as the WSSD with its emphasis of partnerships between business and governmental bodies.

The assessment of sub-national regional development effects from large-scale hydropower projects should accordingly concern not hydropower as a form of water use only. It should also address to what extent a dam becomes multipurpose through integrated water resource management (IWRM), and what upstream and downstream consequences may follow. This often brings international effects to the picture, since many exploited rivers are international.

The potential effect of dam projects for poverty reduction is of special interest for all countries in the developing world. People in the social catchment areas – who rely on the river as a resource – may be the first to be struck by detrimental consequences of a project. They may, however, also make up potential beneficiaries of regional development through improved social service, upgraded regional economy and more job opportunities. These issues have normally been attended more or less as a matter of routine. With the increased pressure from sustainable development requirements, the need to be more specific in the information provided has increased. Issues like what the impacts are for different social categories come in focus. There may be direct and indirect effects to be differentiated. Also specific issues within gender, wealth, and age differentiation may exhibit significant differences. The trend over the past decade has been to put more substance into this kind of documentation. In the process there are also a number of success assumptions that can be challenged later (Chap. 5).

Chapter 2 ended in a discussion on regional development around water resource based projects such as hydropower. As already underscored, hydropower projects are expected not only to produce electricity, but to be multipurpose, and also to have both local and regional "side effects". The local perspective includes irrigation, water supply and sanitation, flood control, and fishing in project planning. The distribution of the water resource in such sub-sectors is expected to be attended and change as a result of hydropower schemes. Also the electric energy distribution is expected to change. Rural electrification should become more accessible, so that both living standards and production capacity increase. Success with rural electrification, usually referred to as a major development target for hydropower and also for poverty reduction (cf. International Hydropower Association 2004:17), becomes a good indicator of regional development generated by hydropower development.

The current chapter accordingly puts focus on rural electrification in order to see its regional development context. This will be done so as to go into detail with the question how poverty reduction in a region may, or may not, emerge due to hydropower projects. This issue should be particularly attended since infrastructure development projects need to be designed and carried out in accordance with the Millennium Goals to reduce poverty to half by the year 2015. Extreme poverty shall be totally eradicated by that date, along with other social goals.

Rural development has, since the early 1970s been a major argument for rural electrification. The effects on regional development have been slow to appear, however, as verified in the current chapter. A lot of momentum was built around rural electrification as a new energy supply through a grid in rural areas. Now it shows that regional development takes place only when rural electrification combines in a region with other sustainable development reinforcing factors; especially market access. The great expectations on rural electrification often heard, as a major regional consequence of hydropower generation, may be reflected in regional development. However, whether this also implies poverty reduction and the impact of energy generation on local consumption are open questions, addressed in the current chapter. Among the critical issues are who the potential electricity consumers might be, and how rural electrification can be pro-poor as a development strategy in a situation where investment is made in hydropower. This particular opportunity for poverty reduction orientation is the context where the current chapter revisits rural electrification projects. It does so with the idea to address the realism in proponents' assumptions about the capacity of rural electrification to lift a region in such a way that it can be registered as collective compensation for hydropower implementation negative local effects.

3.1 Rural Electricity in the Lives of the Poor

Indirect regional effects of energy distribution due to hydropower through rural electrification are claimed to follow on hydropower establishment.[1] In order to specify
key questions, the case of Lesotho is drawn upon for addressing who the poor but
nevertheless potential electricity consumers might be, what their demand for electricity will be, and what factors should be considered for individual households
when deciding whether to connect to the grid or not. Lesotho is that country in
the world with lowest proportion of electricity consumers (together with Ethiopia)
according to World Energy Outlook (2002).

3.1.1 Energy Consumption Among the Poor in Lesotho

Lesotho experiences the globalization phenomena that are well-known in many
countries; increased market dependency, polarisation between rich and poor, and
a growing vulnerability for those with unreliable resources. Typical features of
poverty for Lesotho are increasing unemployment in the South African mines, growing inequality and increased vulnerability among those in the rural population who
find themselves in the lower social quintile.[2]

Signs of improvement can be seen, with a slightly increasing economy in spite
of ceased labour migration to South Africa, but with much of the population below
the formal poverty line. As much as 66 % of the population (2 million) live below
the poverty line of 2 USD per day (about M13) according to the World Energy
Outlook (2002), showing that 5 % of the population has access to electricity. The
monthly poverty line per household is M124 (Wason & Hall 2002:4). The positive
development features according to them include:

- An attitude change about government role; expecting it to be reactive to local
  initiatives rather than pro-active
- A slight decrease in poverty
- Less crowded housing
- Better access to latrines
- Improved access to clean water
- Clearly decreasing household sizes.

Among trends in the statistics behind this set of observations can be noted that employment, including self-employment seems to be on the rise. In particular women
as a category are creative and take initiatives within small-scale sectors. There is a
shift away from farming, and the emerging urban areas could retain a role of growth

---

[1] This section refers to a study for Sida in 2002, where the author contributed the case of Lesotho.
Other countries in the study were Botswana and Ghana. The three studies were later integrated,
and occasional reference is made to the synthesis result.

[2] A poverty assessment in Lesotho is given in Sechaba (2000). See also Wason & Hall (2002).

centers if supplied with proper infrastructure services like electricity. Resettlement under such semi-rural circumstances with lacking basic service structure is a plight for the Poor. However, electricity can be provided more effectively than in a rural situation. Small urban centres have a potential for sustainable development.

A study of household energy consumption suggests how rural electrification has a potential to reach poor people (Sechaba 2000:93). The use of electricity is clearly correlated to income level. This shows in the consumption of different energy sources in low-income strata. For instance, the use for lighting by income strata shows how households with low income use the available light sources (see Fig. 3.1). The poverty line at M124 cuts straight through the next-highest income categories among the poor. This is also where the use of electricity for lighting begins to increase. In this way electricity picks up with income, while paraffin consumption goes down: The extremely poor rely on some firewood but predominantly paraffin and considerable use of candles. As income rises in the figure the use of paraffin decreases, candles increase temporarily, and electricity enters the picture. Next income level (behind the back wall in the figure) sees an increased dependence on electricity for lighting. The figure suggests what is illustrated below – that shifting to electricity for lighting is not so far away economically for the Poor.

The y-axis in both Figs. 3.1 and 3.2 indicate volumes consumed in order to express the trends in consumption pattern.

For cooking, as for lighting, Fig. 3.2 shows a distinct income threshold with shifts between energy sources. Here, electricity tends to go up while firewood goes down (both logs and shrubs). Gas is the most popular source for cooking when income

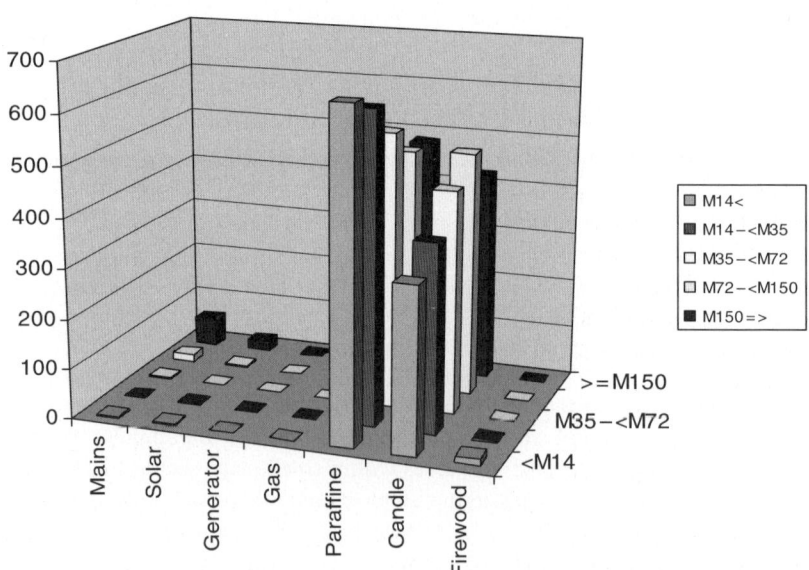

Fig. 3.1 Lighting by economic quintile in Lesotho according to Sechaba (2000:93)

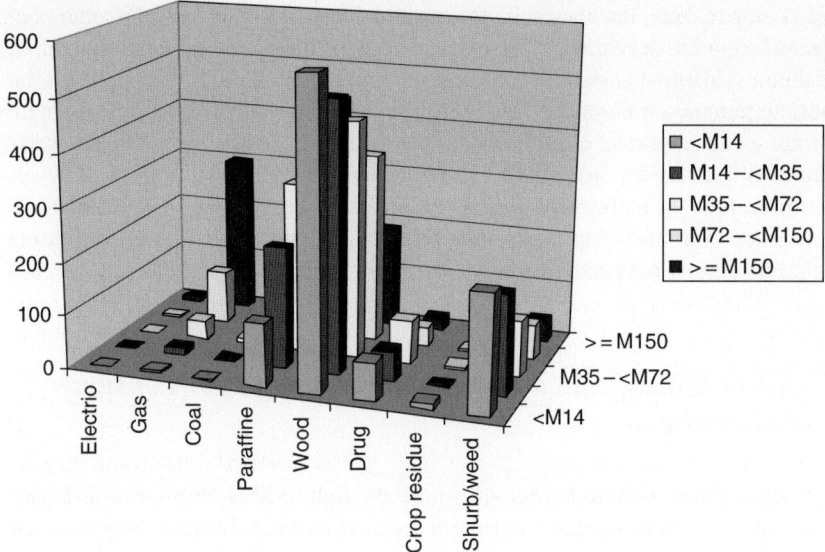

Fig. 3.2 Fuel use by economic quintile in Lesotho (based on Sechaba 2000:98)

increases since there are almost no installation costs. Electricity use increases, as for lighting, when moving into next quintile (the back wall in the figure).

There are environmental implications to be noted. Woods, shrubs and weeds are the exception for cooking; these are "poor person's paraffin". The fact that the study brings out firewood collecting as a significant feature is a poverty sign. In both cases, paraffin and firewood use are an environmental issue. A change of energy source into electricity, when viable, is a step forward, both for cleanliness (health) and natural resources conservation (environment). It links up with poverty and gender considerations for sustainable development.

Rural electrification does not primarily target "pure" rural electrification sites but rather growth centres; like villages or small towns where schools, dispensaries, hospitals, and other public amenities, together with shops and other SMEs, are located. The Poor as electricity consumers and SMEs as both consumers and creating job opportunities for poor people are two subscriber categories of rural electrification that will be investigated a bit further. The rural growth centres have the greatest potential in the country for regional development. Especially district centres host a variety of small businesses. This would be where the expectations from sustainable development policy can be met. Access to reliable and affordable energy is seen as a condition for poverty alleviation by generating jobs and income (WSSD 2002), and the creation of job opportunities after daylight is necessary for increased productivity and hence job availability (MDG, 2000, and a follow-up Plan of Action 2005). A gender side can be added. since the small enterprises are female headed (Orgut 2002). This is not unique; in neighbouring Swaziland for example, 84 % of SMEs are led by women (Marcelle & Jacob 1995).

For policy makers, the answer to the question how rural electrification can contribute to regional development is expected to be found in the combination of these themes; affordable energy, job creation, and gender equity. Judging from the Lesotho experience it could be that its role is hardly as a first-rank development driver but as an important contributor to infrastructure development. Proper water accessibility is given top priority as a prerequisite to development, followed by social service in the form of education and public health. But highest up on the demand list, also social service rather than production oriented, is the need for good lighting at night for security reasons. This will show in the next section.

3.1.2 Rural Electrification Expectations Among Poor Consumers in Lesotho

Demand and expectation to access electricity through rural electrification is linked with affordability. When going into further detail and considering the different costs involved, the Lesotho example suggests a potential to widen the electricity consumer base to include also poor strata of population. The following summary demonstrates the situation (Orgut 2002).

> Lesotho inhabitants are poor, to the extent that private electricity consumption without loans or subsidies is out of the question for many. 68 % of the entire population is classified as "poor" in the Lesotho case; defined as having a monthly cash income of less than M124; about USD18). The lowest quintile has an income less than M57 per month. Recent studies indicate a deteriorating situation in that the poverty struck population has increased in numbers. This shows particularly for the female headed households in two ways; a growing gap between rural and urban women, and the fact that the gap between urban female-headed and male-headed households does not close.
>
> This picture of dramatic poverty at individual and community levels stands in contrast with the Lesotho profile in international statistics. The country compares quite well with other African countries in terms of child health, education, general health and national economic development. Nevertheless, poverty is among the highest. The explanation to this seeming contradiction is found in uneven resource access. The management of subsistence security evolves around an individual's presence in many systems (wage labour, farming, self employment, government schemes, livestock, collecting firewood and other natural resources, informal sector). Such systems of resource aggregates are functional as long as assets can be redistributed. In times of crises they cease to function, even collapse. They form a major alternative to regular wage earning. The current Lesotho development process shows that the proportion of salaried persons has decreased, while those remaining in the sector have seen their life quality improve. The stratification of society is in other words increasing.
>
> Judging from the interviews and discussions carried out within the frame of the background study (Orgut 2002), there was room for improvements in management of rural electrification. The consumers in the study particularly commented on marketing practices. The impression conveyed was that individual costs for connecting, even within the same housing scheme posed a problem, especially since the cost for electricity consumption, primarily for installment, was such that poor strata of the society would not consider connecting, whether prices were fixed or varied. Potential consumers called for reduced prices or payment schemes. Going into consumer detail, however, the interesting pattern emerged that it

was installation costs and not running costs that prevented a potential electricity consumer. It is even so according to the study results, that once installation is paid, energy consumption becomes cheaper with electricity than with other sources.

The description above gives the background for those who are poor with reference to their decision whether or not to shift their energy consumption to electricity. The emerging picture is that rural electricity consumption is out of the question for the poor strata. Attitudes to electrification are positive, although people see no direct connection between rural electrification and reduced poverty. An exception is SME owners; electricity is in great demand with them, and they also see their role to be to bring electrification into productive activities: Poverty can be indirectly combated through job creation due to improved infrastructure. But direct investment in electrification remains impossible for the very poor. The following case shows the situation of a near-poor woman. It suggests that the cost equation solution is not so far away at her economic level (from the author's field-notes, presented in Orgut 2002).

The experience of Ms Rose

The household survey findings from 2002 endorsed the observation that domestic electricity is not conceivable for the "common person". In-depth interviews confirmed this. At the time (2002), Ms Rose's monthly cash income was minimal (between M100 and M200 [USD 17–33]). Her grown-up children supported her by paying the loans for the small cement house she inhabits in a modern housing scheme. The loans had just been paid after a 20-year period. Rose had no means to afford electricity without another 10 years of debt added if she were to become a consumer of electricity.

A typical calculation for a minimum house in this housing scheme on self-help basis that was started 1979 is as follows. Investment in the plot M1,500 [USD 250]; labour costs in total M142 [USD 23.7] (walls, roof, plaster, carpenter) given that own work is invested; materials and other costs (such as fences) on a 20 year loan at M17 [USD3] per month.

This can be compared with an intended electricity scheme with an installation cost at M500, if minimum 50 persons joined, plus a loan for the remainder of the investment at M32 per month for a 24 month period. On top of this cost came the standard charge (said to be M27) and operating consumption (the lowest figure the team saw was M7.75). The aggregated monthly cost with a loan adds to $M(32 + 27 + 8) = M67$ per month. This would be about half the amount of cash available for Rose.

Can Ms Rose afford that? The cost for maize meal is M45–47 per 25 kg. The cost for milk is M2 per half liter. Vegetables can be produced in a housing scheme plot (assuming that no new droughts and concomitant water rationing occur). This means that she on a dramatically tight budget could possibly come close to being able to afford the electricity. But this is an unrealistic calculation. Also other costs occur, such as gas and kerosene, clothes, etc. But also the income side of the budget has a hidden post; remittances sent by children or relatives.

Ms Rose's way of reasoning was in line with this calculation. She had no intention to tie her hands so dramatically even for a two-year period. Still, she was aware that had she once made the investment she would have a realistic chance to meet costs. The relatively new prepaid card system was hailed since it allows the consumer to monitor costs and cut down on consumption in time. Especially after even further investments in appliances Ms Rose's monthly costs for energy and light might even be lower than today. A considerable gain would also be the easiness and cleanliness in comparison with bringing gas and kerosene, and standing the smoke and the smell of the kerosene lamp.

This illustration suggests that rural electrification is not for household consumption among the non-salaried, especially not among the poor and among female-headed households. Ms Rose is typical for many peri-urban inhabitants and potential consumers of rural electrification. Targets for rural electrification as mitigation measures will bring improved social service and indirect benefits for a population (schools, health centers and street lighting/security) and private enterprises (SMEs). Any connection to the electricity grid of poor strata in massive scale is unlikely. The benefit lies instead in the indirect effects. For a regional planner this means that a concentration to some kind of rural growth centers and not purely rural situations is to be prefered.

The quoted study supports the conclusion that it is the indirect effects of rural electrification that can be expected to benefit the rural Poor in a regional perspective (resettlees may make an exception with special treatment). The following main conclusions on electrification and poverty after implementation can be made out of three country studies in Botswana, Ghana and Lesotho (Orgut 2002).

• Life quality improves generally through rural electrification projects, also for the poor strata of population
• Installation costs block direct access to electricity for the Poor. Loan schemes are called for to change this
• Running costs are manageable for poor sections of the communities, possibly with price adjustments
• There are considerable indirect social effects from rural electrification; security, improved education and improved health services rank highest.

Social impacts from electrification are easy to identify. In the Lesotho study, better light at home and cleanliness were prime arguments for electricity, followed by the observation that crime is hampered and security increased. This security aspect was emphasized by low-income women who had no electricity but whose neighbours had, so they could see the benefits.

The economic consequences of electrification are more difficult to document. Potentially rural electrification reaches social groups that set the pace towards a more sustainable development. Even if no direct economic effects will materialize and are not showing on poverty maps, electrification still has development potential through the emergence and growth of SMEs.

3.2 Rural Electrification as Support to Private Enterprises

Social safeguarding, making sure that no one is worse off after a project than before due to project implementation, is a key concern in regional development in general as well as specifically for hydropower project planning. The hydropower project design has to minimize negative impacts. With poverty reduction and long-term sustainability requirements of today, forms for livelihoods come in foci, also including non-farm activities that imply migration and resettlement, and the emergence of small urban centres. The sustainable development demands open for more successful rural electrification when the frame of reference shifts from profitability to regional development. Of particular interest has been the emergence of small enterprises and consequently job opportunities.

Two cases of such small enterprise developments are referred to below. They address the potential for new job opportunities and the ways in which private enterprises may diversify. One is a study over 30 years of one township in Kenya, from its days before electrification until today. This study illustrates the development over time. The other one comes from the study in Lesotho already dealt with above in Sect. 3.1.2 and concerns the effects of rural/urban electrification on female-headed small or even micro scale enterprises. It also looks at the potential for electricity distribution for the electricity company by addressing potential consumers' willingness and capacity to pay.

Direct and indirect purposes and effects are specified. By linking up with the structural change following the implementation of neo-liberal policy and market orientation, and with stakeholders' expectations to meet benefits, the study specifies the potential consequences in affected regions as seen from sub-basin and local community perspectives.

3.2.1 Long-term Effects of Electricity on Business: A Thirty Year Long Case from Kenya

Rural electrification has been debated since the 1970s due to low return on invested capital. The other argument has been that it is a costly but necessary investment in infrastructure for rural development. In the study (Orgut 2002) quoted above in Sect. 3.1.2, comparison is made between three African countries. One observation is that there is a slow increase in the number of added electricity consumers. The study caused a discussion over the uncertainty about the development effects from rural electrification. It can not be taken for granted that rural electrification contributes to rural development in a broad sense. Looking for eradication of poverty as a sustain ability goal in particular, could turn out to be wishful thinking. In order to address long-term effects, the experiences from a township in Kenya, Isiolo, may offer an insight into long-term development; from 1972, prior to electrification, to 2002, 30 years after the electrification of the town. At that time, with the technical Lalander report (1971), there was a fast growing interest in rural electrification as a tool for rural development.

The SME and household economy studies in Isiolo town just prior to the electrification took place in 1972 (Hjort 1974). The demand for electricity among potential consumers at the time is given in Fig. 3.3 below. This study was carried out among all SMEs in Isiolo before electrification and with a sample (about 100 SMEs) in neighbouring Meru town which had been connected to the grid much earlier. The figure shows the very high expectations in Isiolo and the much more low-key observation of what happened in Meru. In Isiolo most respondents expected the distribution to be much more reliable than from diesel generators. Practically all expected

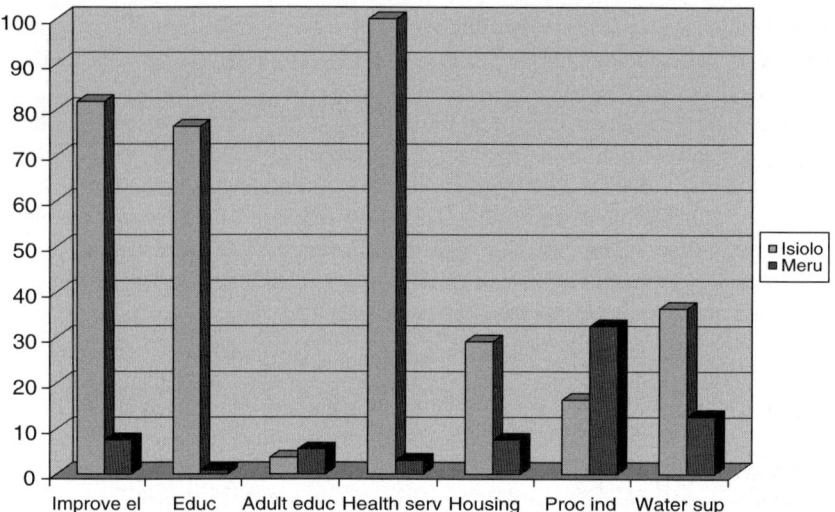

Fig. 3.3 Wanted improvements in two towns (Isiolo, not electrified and Meru, electrified) in 1972 (Source: Hjort 1974:84) [in % of samples]

health service to improve, and most anticipated evening classes, photocopying and other new functions in the schools. The Isiolo expectation was overwhelmingly to see improved social service. Judging from the Meru experience this would not materialize. On the other hand there would be good reason to upgrade expectations on an expanding process industry.

Going into details of job opportunities and business diversification is not possible based on this kind of comparisons only. It becomes too risky because several factors influence; not least the level of regional development when the study was undertaken. In order to achieve more reliable data diversification due to rural electrification, the author therefore carried out a follow-up study in Isiolo 30 years later. Fig. 3.4 shows the results. The number of businesses had increased only marginally; from 96 to 98 during 1972 and 2002. For the same period they had diversified, however; slowly but clearly. Practically all registered businesses use electricity in 2002. The diversification has largely been into such activities that demand access to electricity; hotels, carpentry and garages. No significant effect can be noted on intensification in business operations, reflected in change in the monthly turnover. In 1972 the turnover for retail business was around KES 3,000, and in 2002 KES 35,000.[3]

The Isiolo case study results are in line with the findings in the Orgut study (2002) referred to above, that rural electrification creates improved business opportunities, but that the change is gradual, and the investment in rural electrification has to be paralleled with other infrastructure investment in order to create an enabling

---

[3] A loaf of bread costed KES 1.20 in 1970 and KES 24 in 2005. The whole virtual increase is due to inflation.

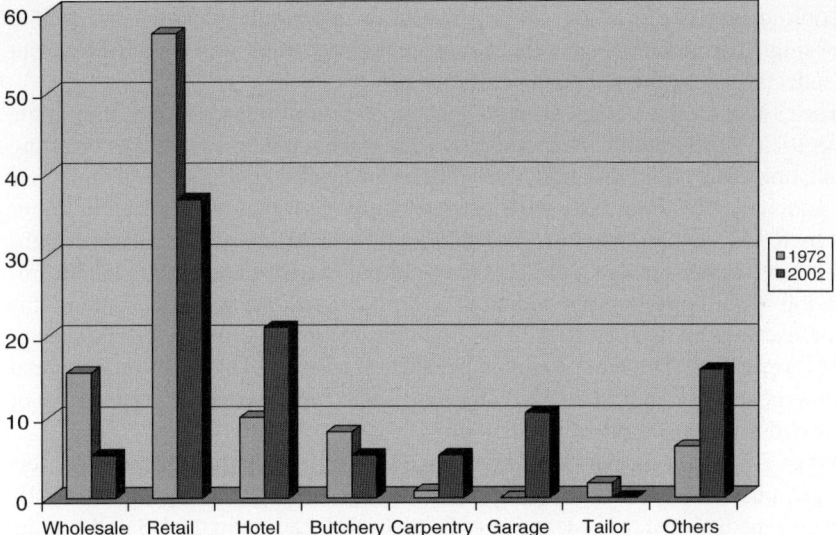

Fig. 3.4 Business diversification 1972 and 2002 as reflected by the number of enterprises per sector; y-axis shows number of businesses (author's field data)

business environment. For hydropower construction to link up with regional development plans, an investment in rural electrification is close at hand. It should then be seen as a mitigation input, contributing towards regional development but requiring also other further investments to be made within the regional planning frame, adding up to an integrated approach towards sustainable development.

3.2.2 Rural Electrification: SMEs Open New Niches for Women's Initiatives: Maseru (Lesotho)

The prime impact of rural electrification on regional development seems to be indirect. In the Lesotho case (Orgut 2002), the electricity company staff anticipated production effects in households and growing small enterprise as the prime targets. The surveys reflected this anticipation nicely in that interviews with both categories, households and enterprises, showed expectations to be high on changes in work or income generation, not least through private enterprise. However, opinions varied in villages that have been electrified some time ago. Those having electricity experienced that little change had actually taken place (over 80 %), while those without electricity in the same villages saw a change around them in terms of upgraded business and job opportunities in 55 % of the interviews. Two conclusions were drawn in the study; that change is long-term and that it is most telling for the poor, whereas the better-off see it as marginal in their way of life; short-term effects are found in social service, not production. Combining with the unusually long-term observation on Isiolo over 30 years above, a safe observation is that rural electricity in itself is not a single driver towards entrepreneurship and large-scale start-up of SMEs. But it is a decisive element in regional development.

The case of Lesotho concerns those just out of poverty traps, and how they relate to electricity consumption. It demonstrates that pricing policy, specifically targeting installation costs rather than tariff subsidies, is the key to mobilizing larger numbers of customers. This empirical observation is supported for a wide range of countries by Komives et al. (2005:6). Provisions of infrastructure (such as electricity) for SMEs in rural areas makes a double contribution towards economic growth by creating job opportunities and by adding to government income through taxation. This is one reason why rural electrification often has been a priority sector in development cooperation. The hindrance that installation costs bear to SMEs in a regional development goal needs to be overcome, and could be one target for a new style of integrated hydropower project mitigation.

When going into the details of emerging SMEs in Lesotho patterns show. A survey of household enterprises with and without electricity, carried out in 2001, showed a median cost for installation of 5,000 Loti (6.18 Loti to the US dollar at the time). About 10 % of the current and potential consumers categorically did not want electricity, while 63 % were clear that they would be ready to pay for the installation once the opportunity comes (or did pay when the opportunity came). Adding to this expenditure was the first connection fee (at 1,500 Loti or above). One third of those without electricity found this additional expenditure impossible, being the factor that prevented them from subscribing. 20 % were willing to pay but not as much as the M1,500 just mentioned. The third expenditure, wiring of the premises, could be covered by half of those currently without electricity if the cost level stays at M500. Since the study showed that the energy cost for those without connection to the grid are just as high as the costs for electricity from the grid, there seems to be an obvious case for a loan scheme.

The Komives et al. (2005) study, suggesting that subsidies in electricity do not particularly reach the Poor, seemingly speaks against such a proposal. Structuring the costs as above, however, tells the story that subsidies are needed but not for consumption. This is the same conclusion in both studies. But the bottle-neck shown above is connection cost. That message might be important for sustainable development, given the insight that affordable electricity is a major building issue for poverty reduction, in accordance with the Millennium Development Goals (2005).

The case of Lesotho is special in that a dramatic decline in wage labour recruitment in the South African mines has meant much unemployment in the local communities. Not only is less cash coming back, but also a generation has developed of unemployed men who remain home permanently. Their profession is wage labour, and this means that they are not used to taking own initiative. This was a situation discussed for instance in Sechaba (2000). When discussing the matter in the deep interviews partly outside the quoted study, it shows that a generation of unemployed men is in need of repatriation into the Lesotho community, especially when reflected against a generation of entrepreneurial women, ready to test new income sources.

This whole issue of home-coming men is of course a sensitive gendered issue. Women interviewed emphasized that the men's minds must change. They also held forth that micro-scale enterprises are typically women initiative. The momentum for change and creativity was to be found in this sector, and it would provide the only job option for the now idle men. This was claimed in a heated debate where the counter-argument was that it would be of no use to try and establish more activities when the return can only be low. Instead there should be targeted government initiatives to provide income generation (Orgut 2002).

The intention with mentioning this debate is to indicate that SMEs have high female representation in Lesotho today. Rural electrification is directly connected both with improving women's situation and with building a capacity to mobilize more men in the sector. Both aspects, household and SME electricity, are gender sensitive in a positive way; helping to address the problems of former miners and to create more jobs in a female dominated sector.

The Lesotho strategy for rural electrification is to connect both households and small businesses to the grid in order to support sustainable development in rural areas. Selected municipalities are linked up step-by-step. Additional services like premises for SMEs to rent are provided by local authorities on long-term rent contracts as part of a strategy. This could lead to growth centres of the kind discussed in Sechaba (2000:205). These are equipped with schools, health centres and other service amenities. If hydropower mitigation should aim at one regional development effort, rural electrification is a good candidate. And this would not only be because of the common denominator electricity. Rural electrification also has the potential to target those who are moving out of a poverty trap, and in particular women.

3.2.3 Rural Electrification and Regional Development Prospects (Botswana, Ghana and Lesotho)

The long-term development in Isiolo, Kenya, presented in Sect. 3.2.1 supports findings from Lesotho (Orgut 2002). The latter also includes studies from Botswana and Ghana. In the cases of electrified SMEs in Botswana, Ghana and Lesotho, the enterprise profiles look as follows (Fig. 3.5):

The figure shows the average number of job opportunities per enterprise for each sector (not all sectors are supplied with electricity in the municipalities studied). It is a picture of micro-enterprises that emerges. With the exception of Ghana, where focus has been on rural villages, attention in rural electrification projects has been given to rural townships or urban suburbs. Rural electrification contributes with a reliable and clean energy source, allowing of course for light, but also stove/fridge and some light machinery (typically one machine per enterprise).

The many household enterprises behind the statistical profiles combine electricity for domestic and for enterprise use. These micro-SMEs are often not oriented towards growth but risk reduction. With this in mind it is noteworthy that the potential subscribers' expectations fall in production oriented effects. The expectations

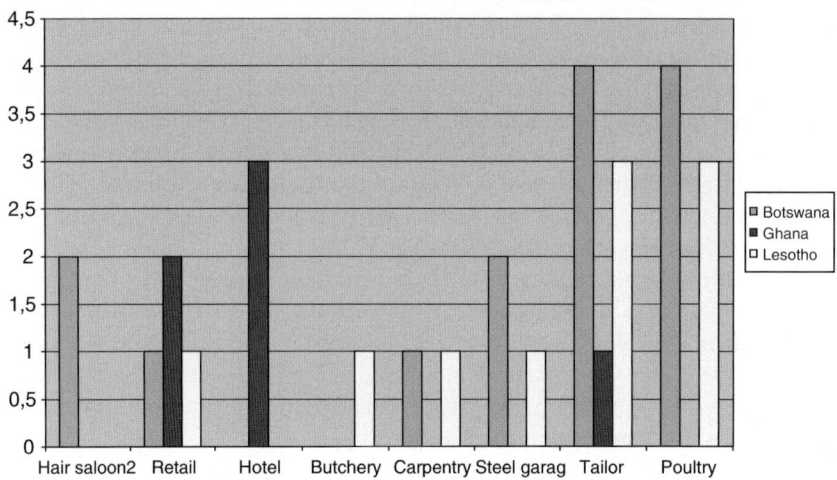

Fig. 3.5 SME sectors in three countries (Botswana, Ghana and Lesotho)

in the three countries on positive effects from rural electrification are (ranked) as follows:

1. More employment
2. Promotion of SMEs
3. Longer work-days
4. Improved security
5. Rapid business growth
6. New types of production.

These expectations contrast to the reality in Isiolo, Kenya of a slow change in SME profiles. In both cases there is a hope for job opportunities. Going more specifically into the micro-scale enterprises, where household and enterprise live in symbiosis,[4] gives a detailed picture. Energy costs represent typically 10 % of the household budget. Subscribers are not the real poor, but people living in permanent houses, also with some sideline business. Improved lighting is for them by far the most significant improvement, with radio/TV and refrigeration as followers up. In a majority of cases householders would also obtain equipment with business potential (cold drinks, food selling). Very high on the demand scale are also improved education (photocopying and evening classes) and health (improved services due to refrigeration).

The likely conclusion is a rift between the better-off and the more permanently poor in rural areas, where the better-off look for higher living standards, the poor

---

[4] Micro enterprises are those "small-scale enterprises" (SSEs) that operate with low value products. Otherwise they have the same features as for small enterprises; "management lies in the hands of one or two persons who are also responsible for the major decisions" (Harper, 1984:5 quoted by Kugbei & Turner (2000:10).

for job opportunities. Regional mitigation through rural electrification might in in-stances be a suitable approach towards sustainable development by meeting both demands. It must then be kept in mind that the regional development effects will not come automatically but only as a result of integrated planning that also com-bines education, health and credit systems. Next section therefore opens up towards a broader regional view on integration where focus is on the vulnerable situations.

3.3 An Integrated Approach to Hydropower Dam Consequences in Regional Development

The last two sub-chapters have addressed two regional conditionalities relating to rural electrification: How accessible it is for the local populations to benefit from improved social services, and how access to electricity is made use of by SMEs for production and job opportunities creation. Rural electrification is claimed to be a prime result from hydropower planning:

> "Hydropower schemes have the ability to significantly reduce poverty and enhance quality of life in the communities they serve. Access to electricity promotes new economic activity, empowers women by reducing domestic and repetitive chores such as firewood collection, improves health and education services, and provides a cleaner and healthier home environment. Hydropower infrastructure, such as reservoirs, also provides multiple-use benefits, particularly through increased availability, reliability and quality of fresh water supplies and reduces flood risks." (International Hydropower Association 2004:17).

A main observation has been that effects are limited and long-term if not com-bined with other inputs. The current section opens up for this broader issue for rural electrification and its role in regional development. In order to do this it touches on water-related experiences from integrated regional approaches. It links back to the National Hydropower Plan Study in Vietnam, the NHP Study, and its rationale to integrate social issues with environmental, economic and technical ones. The methodology is given in Sect. 1.6.[5]

Potential for hydropower projects appears in upland or highland Vietnam. Key is-sues for the development in these areas are poverty, ethnicity and the multipurpose use of limited water resources. They connect with the 20 focused social assessment issues for the NHP Study Stage 2, presented above. Focus is on poverty, how to reduce it from within the water sector (where hydropower belongs), ethnic minority situations (a majority issue for hydropower reservoirs' location), and integrated wa-ter resource management (through multipurpose hydropower project orientations). A foundation is laid for monitoring poverty reduction with this selection of indi-cators of social effects. Reflections on the key question of the real potential of hy-dropower projects for poverty reduction can be documented.

There are issues for regional development at two levels; passive effects and dy-namic measures. One set of issues may concern effects such as who benefits, if

[5] This section draws on the overview report in the draft Phase III report (NHP Study Stage 2, SWECO 2007).

upgrading really benefits the ethnic minorities, etc. Will the effects follow the same pattern as with poverty reduction figures in Vietnam, towards "poverty pockets" in the midst of a generally effective poverty reduction? Another set of issues concern what kind of development hydropower projects bring, who is in the position to benefit from new opportunities, including migration issues.

3.3.1 Poverty Reduction and Potential Hydropower Effects in Vietnam

Poverty reduction is an over-arching goal in Vietnamese policy thinking. The national strategy on poverty states: "Poverty reduction is not only a basic social policy that is accorded special attention from the Government of Vietnam, but also an important component of the development objective" (Vietnam 2002:1). Interpretation of the Comprehensive Poverty Reduction and Growth Strategy (CPRGS) for the water sector means that one of its objectives is to reduce poverty. National government donors like Norad and Sida, as well as multinational donors, such as the Asian Development Bank (ADB), also have poverty reduction as an overarching goal. Poverty can be eliminated through public policy and action (Vietnam Development Report 2004). The donor trend today is to seek more decentralized participation profiles by forming effective partnerships with governments and with a whole range of stakeholders involved, from within each country (Dedolph 2002, interviewing Cedric Saldanha of the ADB). Alleviating rural poverty is today's dominating proposition in development policy world-wide. The message reflected in the WSSD, to combine the three legs of sustainable development, is well taken. The CPRGS parallels economic growth with poverty reduction (Vietnam 2002:1-2). It also goes further to indicate a strategy to harmonize economic growth with social problem solving:

> "The tasks and objectives contained in the poverty reduction and growth strategy not only call for targeted measures to support specific poverty groups, but also see linkages within the matrix of policies that include from macroeconomic policies, policy on structural adjustment, sectoral development policies and measures to social safety policies of all sectors and levels that must work in tandem to ensure sustainable development" (*ibid*: 3).

The water sector generally, with hydropower as a sub-sector, is invariably considered to be a key to reducing poverty. It is closely related both to natural resource management and economic performance of a country's regions as well as local communities. Regional planning in Vietnam has to face features for poverty reduction, such as a large proportion of the population below a poverty line (though most out of extreme poverty), skewed social stratification and uneven spatial distribution. While "poverty pockets" do not appear spatially, the tendency is to find most poverty both in rural areas and in urban areas with underemployed popuations.

The concentration of permanent poverty is high in remote, isolated and mountainous areas. Ethnic minority groups exhibit extremely high poverty rates that until today do not decrease at the national average rate. Yet, the country is generally successful with its poverty reduction. A number of causes are given for this state of affairs (CPRGS 2002): Competence building, job opportunities, improved basic services (safe water, environmental sanitation, water drainage, lighting and garbage collection) as well as social services (health care and education) and migration (unregistered migrants have occurred in unknown numbers in the wake of structural adjustment of the state sector) are listed as priority measures to reduce poverty. The problems are there: Agricultural work brings highly unstable income in return (fluctuations strike hard); limited natural resources affect not only income but also force poor people to live in areas susceptible to natural disasters (typhoons, floods, drought); underdeveloped infrastructure of poor regions (roads, education, health, safe water, electricity, small-scale irrigation works, and markets). The listing of water sub-sectors in the report has been made through a content analysis as described in next section.

All these are elements in a sustainable regional or river basin development planning. Typical to a situation featuring a vicious cycle of poverty, is a general lack of resources. Household vulnerability is high since a large proportion of the population lies just above the poverty line. When bringing poverty considerations into the water sector, irrigated agriculture and drainage are obviously included against the general background of current vulnerability for farmers; flood control and disaster mitigation rank high, safe water and health/sanitation problems appear in the infrastructure problem complex, and hydropower ranks high for electricity production. All these sectors connect to food production, health and mobility (infrastructure, communications), issues that are contributing to the level of empowerment of poverty households. One key question for mitigation is who participates in the selection of priorities. The issue is no longer technical but stakeholder driven, as already dealt with in earlier chapters. But who are these stakeholders, what "mandate" do they have (whom do they represent)? This is discussed in Chap. 4.

Many of the ethnic minority groups in Vietnam are traditionally highland and mountain dwellers. Even though many ethnic minority people have settled down at lower elevations as a consequence of the governmental fixed settlement program, the most remote highland villages are inhabited by isolated ethnic minority populations. The situation of ethnic minorities has been given much political attention over the past decade. This is not only a Vietnamese phenomenon, even though ethnic minorities are highly prioritized in various policy agendas in the country. For hydropower development the issue of ethnic minorities has reached a top priority on the agenda (World Commission on Dams 2000). In the NHP Study ethnic minority issues were given special emphasis not only as a category in itself but also through the way key issues in the social study were integrated with the other aspects of production and economy, naturally, but also of environment.

The linkage between poverty and ethnic minority status is also related to residence in that many of the small ethnic minority groups are local, living in certain districts in concentrated populations. In the case of Song Bung 4, one hydropower project that is currently designed in the country, the resettlement affected population is small by hydropower standards, but it affects a large portion of a small minority group that is quite small in Vietnam. The socio-cultural consequences can be huge, so that social safeguarding may still become a major issue.

The circumstances for ethnic minorities are reflected in the national statistics, they which show great success with poverty reduction with one important exception: ethnic minority groups (one sixth of the population). The connection between ethnic group membership and poverty is obvious. This link is probably strong in Vietnam in comparison with other countries. In general, 50 % or more of the ethnic minority groups in the NHP Study areas fall below the official poverty line. Many studies, such as the CPRGS (2002) demonstrate this fact. Poverty reduction is an issue of particular importance among ethnic minorities, not only since poverty incidence is high but also since living in poverty seems to become self-perpetuating. In situations of resettlement, unavoidable in most hydropower projects, ethnic identification proves to be of particular significance. A project has to address forms for forced resettlement due to hydropower development.

As already argued, resettlement needs to be integrated nationally in regional development plans which include also host populations as well as indirectly affected. In a national perspective the large-scale migration streams go to new economic zones and to urban areas. Cultural distance is an important factor for migration. Since resettlement and social unrest are so decisive for success much attention in the NHP Study was given to these aspects. The highly significant seasonal urban migration was, however, not addressed in the Project.

Beyond this example – globally – lies a key political issue to maintain political stability. This is considered to be achieved through proper mitigation. The reverse reasoning is also applicable; since there is good reason to maintain political stability, exposed minorities find an argument for good quality mitigation.

3.3.2 Hydropower Projects for Rural Development

So far, the concern has been with the indirect effects from hydropower projects on poverty reduction in a regional perspective. The other side is to see hydropower projects as instruments in rural development. Linking poverty reduction and water resources needs a set of indicators in order to formulate a profile. The regional planner who aims to establish a water poverty index (Sullivan, 2001:472), must combine welfare variables and a set of infrastructure indicators (access, time and effort to get water, institutional constraints). On top of these considerations come water management issues and the potential to reduce a water poverty gap through interventions. Can hydropower projects be turned into an instrument for such purposes?

The case of Vietnam is one where integrated water resource management has been linked with poverty reduction. The notion of water poverty has been introduced. A water and poverty index for this proves to be most suited at province level. Soussan (2003) suggests ten parameters; food poverty, overall poverty, clean water access, latrine access, child malnutrition, infant mortality rate, water-borne diseases, poor households in agriculture, ditto livestock/silviculture/aquaculture, and vulnerability to natural disasters. These are also among the key issues for the social assessment of hydropower. An immediate first conclusion based on the water relations to poverty, when addressing water resource management as a tool to eradicate or reduce poverty is that interventions of multi-purpose character are to be preferred. Regional IWRM on river basin or sub-basin basis is well established. Hydropower forms part of river basin management, especially since several projects may form a cascade of undertakings in a basin.

The national priorities as expressed in the CPRGS (2002) with regards to water issues in poverty reduction can be seen in the frequency different water sub-sectors appear in this strategy document. A count made by the author shows that water supply and sanitation, irrigated agriculture and drainage, biodiversity, and flood control are mostly mentioned. River transport, water pollution, hydropower and forestry appear less relevant to poverty reduction. The generic categories Awareness and Management and planning receive little attention. Gender, in contrast, is very much emphasized. Hydropower is not listed as an own sector. However, as argued in previous sections, its design should be multipurpose, and the above four water sub-sectors are also connecting to multipurpose hydropower development.

As seen from the studied projects there is potential in integrated regional approaches hosting back-up planning to hydropower, aiming at job creation and rural electrification. The planning horizon is long-term, and the complexity is considerable. Throughout the studied projects the ambition level has been set to minimize negative social effects. Note has also been taken of positive effects, but in no instance hydropower projects have been seen as tools for other investment planning towards reducing poverty or risk. This includes stakeholders from among local people, when involved in the study; their demands have gone to other planning sectors than hydropower.

The common anticipation at both community and regional levels in the NHP Study stakeholder workshops (SWECO et al. 2004:II:2.4) is that hydropower is a national concern, in contrast to other water sub-sectors that are considered province (equivalent) issues. This split on the demand side suggests a management need for sustainable development to link up administrative scales and sectors more effectively towards regional planning. In this way mitigation measures can reach beyond social compensation (health, education, infrastructure) into production and income generation (also infrastructure; small enterprises, marketing). But that calls for a strategy built on the WSSD emphasis of partnership in which local populations and enterprises could join with local, regional and national administrations in order to build consensus about which the key issues are, where they are located, what priority orders should be, and financing of development inputs.

3.4 Hydropower Between Risk Management and Poverty Reduction

In popular debate, hydropower is generally associated with high social and environmental risk, and nothing in this study contradicts the risk picture. The sets of recommendations, following the global conventions, about how to manage the risks generally are also applicable to hydropower planning. The political conditionality is therefore also to address how hydropower projects can contribute to poverty reduction. The last two sections have given the two possible perspectives; hydropower appearing in a regional context with consequences to be addressed, and hydropower placed into it with the intention to generate desired effects.

A few conclusions can be drawn for sustainable development. One is that the expectation to see automatic development effects from mitigation are probably false. Mitigation can compensate for suffering but not generate direct alternative living for the affected. Furthermore, the programmatic expectation that mitigation through rural electrification as infrastructure improvement automatically brings sustainable development is also false. And thirdly, the popular view in affected regions that hydropower projects are an alien element reinforces their operation within a national frame but alien to regional development.

Some measures seem called for from the sustainable development point of view. One is that the trend to better involve stakeholders has to be extended. Next chapter deals with that. Another is that hydropower projects will continue to operate in isolation and hardly contributing to sustainable development in spite of environmental conventions unless they are linked with other measures; notably loan schemes (such as for connection costs to a grid), SME development for creation of job opportunities, and "soft measures" within capacity building and awareness about resources management as well as regional planning.

Chapter 4
Decentralisation and a Rights Approach in Hydropower Development

The plight of the project affected people, PAP, in hydropower projects is an often sad story. It has contributed significantly to hostile reactions to large-scale hydropower projects. Some hydropower projects have reputedly become economically viable at the cost of a suffering population of resettled people. Compensation has not been sufficient or properly designed. One incidence from Zimbabwe and a study there, on the Bulawayo Water Supply Project in 2000 illustrates a – unfortunately not unique – lack of integration between line ministries. When surveying a village that would need to be resettled there was a sad atmosphere and a question raised: "Why, move again?" To planners surprise it showed that the village already had been resettled about a generation ago due to the previous Cabora Bassa hydropower project.

This small example illustrates the need to integrate hydropower projects into regional development planning. It is just one of the many examples of a lack of continuity and respect for the problems of resettled people living under poverty circumstances. Projects can become sustainable when all stakeholders are mobilized. Hydropower projects still lack genuine stakeholder involvement, notably through influence of the project-affected people. This study has demonstrated how the policy development towards sustainable development has gradually brought stakeholders from context into project planning. There is a growing momentum for a rights approach in development, which is reaching also hydropower. This is what the current chapter focuses on. It is about how different potential models for stakeholding can lead towards improved empowerment rights to involve in specific hydropower related issues, such as resettlement implementation or downstream effects.

The issue of resettlement is not only littered with past failures. Hydropower project budgets have hardly reached the level of factual costs for resettlement. Large numbers of people have been left to suffer due to problems such as under-dimensioned budgets, administrative shortcomings and limited stakeholder awareness about legal rights. Some reports mention that it takes at least one generation for a resettled population to recover. In spite of global conventions, there are different views between, for instance the World Bank and the Asian Development Bank policies on one hand, and country policies, such as that of Vietnam, on the other, over how much responsibility should be taken by the state over time and for the project-affected people (in particular concerning people who are partly affected).

Decentralization and empowerment to manage mitigation measures are finding their forms locally through the generic emphasis on stakeholder participation. In this way experience from past, and current, hydropower projects can be seen from the directly affected peoples' viewpoints. Suggestions about the most important issues that relate to sustainable development will come forth. True, stakeholder participation is in many instances mere prescription, treated in implementation as a design to follow a check-list *nomenklatura*. But already paying attention to stakeholders opens for questions about the rights of affected people – in particular of course the principle that no one should loose in life quality from a project implementation. Likewise, from the regional viewpoint, the requirement to link up with sustainable development brings economic growth together with both livelihoods and environmental issues. Chap. 3 touched on this from the poverty viewpoint. The change in emphasis in hydropower planning is also relevant for all inhabitants in an affected river basin. The widened information base for discussing issues and solutions naturally must be specific for the areas where consequences appear. This means that the affected region is defined by a river basin's features rather than by a country's administrative structure. Stakeholder mobilization towards regional development would for hydropower planning involve inhabitants from a river basin. Several communes or districts (equivalent) may find themselves partly involved in the hydropower development.

Two perspectives are brought up in the current chapter; the local one of managing life with hydropower project effects, and the regional one of involving inhabitants' knowledge and competence as a way to build long-term sustainable development. Stakeholder involvement becomes different depending on scale. Two cases from Vietnam provide experiences of this. One is derived from local experience; setting up a model approach comes at three hydropower stations. The base for this project has been Song Hinh, a private hydropower station where staff was involved in training interaction with resettled populations.[1] Another concerns river basin regional development and connects with the stakeholder mobilization for the NHP Study's participatory identification of issues and solutions in a river basin perspective. In both cases stakeholders are involved in a process towards mitigation and sustainable development.

The Song Hinh training was carried out together with staff from another two stations (Thac Mo and Ya Ly). Together with the three resettlement populations the participants developed the use of indicators in such a way that it allowed for a model both for early warnings about high-risk development, and for targeted low-cost monitoring of socio-economic and health development (SWECO 2001). The result could be applied widely in the country for the involvement by local stakeholders, project affected people as well as staff on hydropower stations, to further sustainable development. This application has not yet materialized. Under other resettlement circumstances, though, the approach is already applied. A recent example is when the Nam Puoi village in Yen Bai province needed to move. The village had requested

---

[1] This study for EVN should not be mixed with a resettlement study for Sida by Lindskog & Long (2004).

to be resettled because of environmental hazards. The decision-making process that followed (Hjort-af-Ornas and Ngoc, P.T.B. 2007) took the villagers step by step to a preferred solution. Basic obstacles were avoided (site selection, land right issues, domestic water supply, village layout, graves relocation, infrastructure provision and project monitoring by the local community). The interaction between technical staff with their issue competence, and a population with their local competence, became possible through facilitation. In the end decision-making merged stakeholders' positions into win/win for both a project's technical sustainability and that of local community development. Increased trust and capacity building followed.

The second perspective, stakeholding in a regional context, is offered below through the experience from river basin workshops within the NHP Study. Careful stakeholder analyses had been carried out in order to involve all categories of stakeholders. The result for the attended river basins was also assessed as an effort to comprehend how representative involved stakeholders had been. Thorough stakeholder analyses, following the UNCED prescription to involve all categories, were carried out in preparation for each river basin workshop, so that the different stakeholder categories were represented. One group of stakeholders was established for each project affected river basin. Workshop participants carried out SWOT analyses for every possible project site.

Recruitment of stakeholders proved to be so sensitive that the original design for selection was replaced with a top-down way of mobilization. This procedure leads to an issue with sustainable development; how to build confidence in a stakeholder process. Without convincing facilitation several parties might become defensive, feeling that they risk to loose control. Authorities could worry over expensive recommendations and over decreased influence in a politically sensitive situation. For example, local populations might worry over being utilized for detrimental decisions against the background of poor track records for hydropower. Various local and international pressure groups would express concern with hydropower for energy production in the first place, since it harnesses rivers through flood regulation.

The policy declarations on harmonizing growth and equity, on decentralization, and on transparency in decision-making have become tested in implementation. The two cases presented below where hydropower and administration staff have interacted with stakeholders suggest what implementation may look like under hydropower circumstances. The first example is local, involving staff from three hydropower stations designing an assessment model through interaction with resettlees. Of interest here is both the design process and the output in terms of a model and its applications. The second example is regional. Stakeholders from all categories in a river basin have assessed the pros and cons with the establishment of hydropower stations in their river basin.

Attention in this study is on the process of interaction, just like in the local case. An assessment model is applied to analyze the power exercised by the stakeholders. It concerns stakeholders' perceived and real power for decision-making (Stakeholder Participation Model, SPM). The stakeholder power analysis is done as a check on the efficiency in involvement.

Drawing on these cases, a local and a regional one, the issue of a rights-based approach to hydropower projects is touched on as a way of concluding the experiences.

There is a trend towards increased awareness as a result of an interaction process in the first, local, case of three hydropower stations. At regional, river basin, level as well, the exchanges of issues and possibilities mobilizes concerned people in decision making processes.

4.1 Stakeholding Model (1): Local Power Station Staff Interacting with Resettlees[2]

The first model for stakeholding was derived from training of local staff at three hydropower stations. One task was to develop means to comprehend the life situations for resettled populations at their respective work place. The output was a model for rapid screening of such situations in order to understand if compensation had been effective. This issue was raised during operation phases, and staff discovered that lasting or new issues raised by resettlees were hands-on. Both sides, staff and villagers, gained insight into everyday life situations and a deepened understanding of living conditions and constraints put up by the hydropower station operation.

The issue about compensating affected people has been one problematic concern. Even if there is a general agreement that compensation should provide suitable income generation opportunities, it still often occurs that compensation is limited to payment in cash, or providing seeds, over a couple of agricultural cycles. New livelihood activities are normally intended, given the structural changes caused. Hydropower projects are therefore generally multi-purpose as a response to the new demands, meaning that they may be oriented also towards irrigation, fishing and/or transport.

Nevertheless there is a tendency for livelihood situations to remain imbalanced over many years. There are several reasons for this, also beyond insufficient compensation levels or periods. Examples from the three hydropower stations include basic design mistakes such as false assumptions about water availability and terrain for irrigation in the case of Song Hinh. Such mistakes from the drawing board can be mitigated through an early interaction between technical and village stakeholders in the process. The Nam Puoi facilitation just outlined could serve as an example of how a process can be built.

This is the background to the training carried out at the Song Hinh hydropower station. It had been built with Sida support on the assumption that it would be multi-purpose and pay solid attention to the plight of resettled populations. Thorough impact assessments have therefore been done during several years after implementation. These assessments have shown obstacles with the fishing and irrigation goals of the project. As a follow-up, also reaching into health issues, the government requested a training course for staff at three stations plus EVN[3] central staff in dealing

---

[2] The example draws on work done together with TiiaRiitta Granfelt (ENS Consult) for SWECO and EVN in a training course January 2002–February 2003 in socio-economic and health assessment; Song Hinh multipurpose training program.

[3] Electricity of Vietnam, the national power company that is initially operated the Song Hinh hydropower plant.

with socio-economic and health issues among resettlements. This led to the joint development of a method to both monitor such vulnerable situations as experienced by the resettlees, and to come up with early warnings when circumstances tend to become unbalanced. With the down-top involvement the selection of key indicators for screening could be assessed at an early stage. The intention was to introduce a highly cost efficient technique through stakeholder interaction, also with the capacity to signal needs for proper monitoring. The method had the important side effect that resettlees and hydropower staff were able to cooperate in real life and not only in special occasions.

The training course was carried out in 2002–2003. Participants were (technical) staff from Song Hinh, Thac Mo and Ya Ly hydropower plants. The goal was to sensitize and build capacity among local hydropower staff to deal with socio-economic and health aspects among the resettled populations. The intention with the project was also to train local staff at the hydropower stations. One special aim was also to motivate staff to gain insight into the plight of resettlees by involving themselves in the assessment of the living situations of the resettled households. For this end the project set up a process of learning-by-doing. The technical staff should address social issues, not to come up with new findings but to gain insights through assessment design.

A model was developed accordingly, suitable to be applied for both monitoring and catching sudden unforeseen developments (surprise events). The approach was established during the course, through cooperation between participants and consultants. The work became an eye-opener for participants in that their distrust against involving local people vanished. By inviting hydropower staff to design and apply an assessment model among three resettled populations in the vicinity of three hydropower stations, they – as stakeholders – became aware that local knowledge and experience can be vital to the smooth operation of the hydropower plant.

A model that carries out rapid screening naturally cannot replace regular surveys. In this case the aim and capacity are to address vulnerability. The strength, on top of setting up an open interaction between stakeholders, was that it in actual fact led to a consensus building about finding a way to establish an early warning. This warning gives the alarm when in-depth surveys are required in planning and operation of the hydropower site.

The output from the project can be seen to indicate three different livelihood situations in Song Hinh, Thac Mo and Ya Ly resettlement areas in terms of a baseline description for each through indicators, and in terms of vulnerabilities. The following two sections deal with these two aspects since they connect with how sustainable development can be gained.

4.1.1 Monitoring Socio-Economy and Health

The requested social and health monitoring project was built on an interactive process between participants, national and international consultants. The work process stretched over a one-year period, with a design to go step-by-step towards formulating an approach to monitoring that can be tailor-made for each particular situation. In the course of this interactive process capacity of all participants improved their capacity, so that the design of the monitoring model became increasingly sophisticated. In the following account, attention is on the model rather than on the stakeholder involvement in its design.

Three rounds of field training were carried out in the vicinity of the working place of each of the participants in the respective resettlement areas of Song Hinh, Thac Mo and Ya Ly hydropower plants. This led to an anticipated local capacity building. It also opened for a comparative perspective by drawing on three hydropower station situations. Beyond training, the purpose was to develop a low-budget screening model that could both back up regular costly surveys and have the capacity to signal when more in-depth surveys are needed. By focusing attention to a limited number of critical parameters, through formalized ways of observing levels and changes, situations in the realm of socio-economy and health could be followed in a cost-efficient way. The output was one model for monitoring and another one for Surprising Futures (see next section).

Two sets of parameters targeting different localities (project area and village level) were worked upon; first a long list and then gradually building consensus about which items to exclude without loosing too much information. This selection process was a major part of the training, built around the practicalities in data collection.

Data was assessed and organized accordingly at project area level and at village level. The project area parameters (PA1-7, see Table 4.1 below) dealt with the

Table 4.1 Parameter scores for Song Hinh

Song Hinh	Score	Monitor Σ	Social weight	Social score	Prod weight	Prod score
PA 1. Host area relations	1×2	2	2	4	0,5	1
PA 2. Migrants	1×1	1	2	2	0,5	0,5
PA 3. Ethnicity	3×3	9	2	18	0,5	4,5
PA 4. Income level	3×3	9	2	18	2	18
PA 5. Farm output	2×2	4	0,5	2	2	8
PA 6. Education	2×3	6	0,5	3	0,5	3
PA 7. Use of electricity	2×1	2	0,5	1	0,5	1
V1. Project related diseases	1×1	1	0,5	0,5	0,5	0,5
V2. Water and sanitation	3×3	9	0,5	3,5	0,5	4,5
V3. Health services	3×3	9	0,5	4,5	0,5	4,5
V4. Education	3×2	6	0,5	3	0,5	3
V5. Culture conservation	2×1	2	0,5	1	0,5	1
V6. Extension services	2×3	6	0,5	3	0,5	3
V7. Land access	2×3	6	2	12	2	12
V8. Farm output	2×3	6	0,5	3	2	12
V9. Secure access to food	3×3	9	2	18	0,5	4,5
V10. Income structure	3×2	6	0.5	3	0,5	3
V11. Water for cultivation	3×3	9	0,5	4,5	2	18
V12. Nat res exploit	1×3	3	0,5	1,5	2	6

communes directly affected by each hydropower project, focused on potential problem identification in their development context as perceived by stakeholders. The village level parameters (V1-12 below) concerned monitoring village level development, also pointing out the most problematic concerns in resettlees' lives. Local staff made the selection of the study villages, two per hydropower project, so that one village was a project resettlement village and the other a spontaneously resettled village.

In the following step data assessment was based on applying a combined score with quality and quantity considerations for each parameter. The approach was similar to the one in the National Hydropower Plan Study, NHP, presented in Sect. 1.6.

Figs. 4.1 and 4.2 give the final results for comparison of the baseline results for the three hydropower plant areas of Song Hinh, Thac Mo and Ya Ly. The figures are constructed so that the further away from the centre a parameter scores, the bigger is the concern over the related issue. Although the prime aim has been to establish the three baselines, the approach also allows for comparison between the three situations. The visual impression shows that Song Hinh is the most problematic of the three projects both for the project area situations (the directly affected area around the reservoir) and the two resettlement villages in particular.

The figure representation gives a strong visual impression of the situations at the different sites and allows for an ocular comparison between them. The need for baseline studies remains basic in that it provides a full picture. A topic like institutional capacity to coordinate all aspects of a project, for example, does not show in the screening since the indicators can not target this kind of information. Yet, of course, it is a crucial concern. But it has not been pointed out by stakeholders. They are knowledgeable about their issues with a project, but they might not imagine

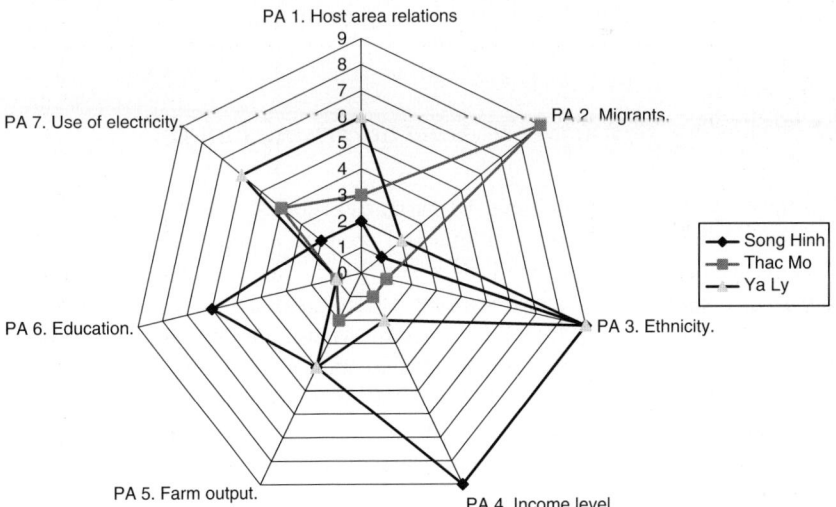

Fig. 4.1 The monitoring model results for the project areas of Thac Mo, Ya Ly and Song Hinh

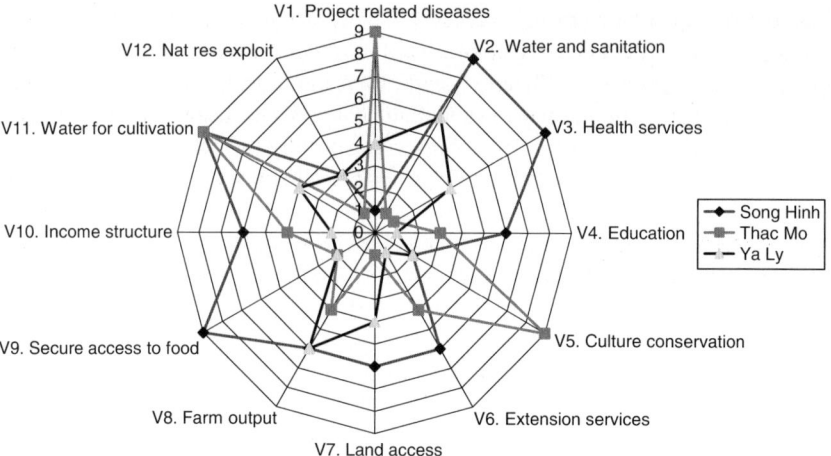

Fig. 4.2 The monitoring model results for the selected villages at Thac Mo, Ya Ly and Song Hinh

mitigation options beyond their experience. The same concern is emphasized by Garcia et al. (2005). They make the point that in the case of Kali Gandaki A in Nepal 1 % of net revenues automatically go to rural electrification as a form for mitigation. The local stakeholders there would hardly have been able to express such a demand without proper awareness raising.

The model approach can also be taken one step further to deal with potential threats based on local experience. For addressing Surprising Futures model, different weights can be ascribed to various crucial aspects of socio-economic and health related issues in order to make clearer the emphasis on production/economy or on social/cultural sustainability. This is done in the next section.

4.1.2 Screening Techniques to Avoid Surprising Futures

One example is given below in Table 4.1 parameter scoring. This table shows both the baseline values and the socio-cultural and production/economic emphasis for Song Hinh. It illustrates the model design: The Score column shows the combined quality times quantity assessments (See Sect. 1.6 for qualitative and quantitative assessment), brought into the Monitor sum column. This column forms the baseline profile. Different weights are then ascribed to indicators singled out to express social concerns or production concerns, as is done in the table (Social weight and Production weight columns). To further assess the Social score, the original Monitoring score is multiplied with the Social weight value of respective parameter. The same procedure is repeated for each parameter, also for assessing the Production score.

Behind the base scores is a long process of focusing the indicators, selecting one-dimensional ones, identifying evaluation criteria for each indicator, ascribing relative values for each parameter with reference to every hydropower station, and also to both quality and quantity values. Methodology details are not accounted for here since the intention is to illustrate an approach. The achieved results can be visualized effectively and form a basis for site discussions among both staff and resettlees. Fig. 4.3, for instance, shows the different outcomes in the case of Song Hinh for focusing on socio-culture and on production both at project area level and at village level. Results may vary with perspective. The project area of Song Hinh is apparently poor whichever approach is used. The social picture suggests considerable ethnic tension. Education and migrants into the area, for instance, poses a

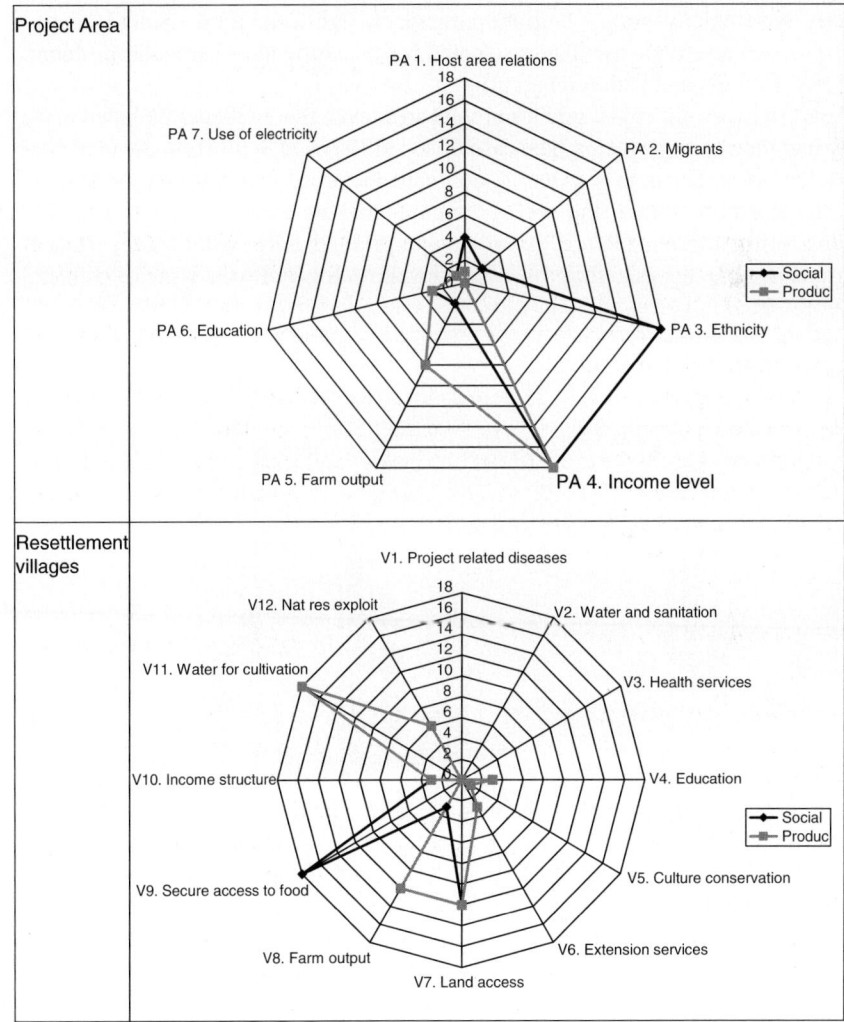

Fig. 4.3 The socio-cultural and production profiles in the Song Hinh project area and in resettlement villages

limited problem only. Going more detailed, into the two villages selected, access to land is a considerable issue both in the social perspective and in the production one. From the social viewpoint food security is the real big issue. This is reflected from the production point of view that water shortage for irrigation agriculture is a fundamental concern. Taken together these factors immediately have a message about high levels of poverty and problems for the project to meet the original policy goal to be multipurpose with fishing and irrigated agriculture as key spin-off activities.

By changing emphasis in this way, through ascribing different weights, a socio-cultural and a production/economic profile can be established for livelihood circumstances in each case as just shown in Fig. 4.4 for Song Hinh. The visual impression from high scores would be a signal of vulnerability. Special concern should be with poverty and ethnicity issues – both are particularly significant for the Song Hinh station. The two resettlement villages selected for the study have particular problems with water access and with food security.

Apart from what was just said for project area level, those villages included in the study had their own problems. The case of social issues at Song Hinh demonstrates how ethnicity and income level totally dominate the problem picture at project level. At village level food and land access dominate and are major problems. Thac Mo exhibits a very different picture; the only major problem felt at village level is that of migrants just like the case for project area level. As for Ya Ly the issue of ethnicity

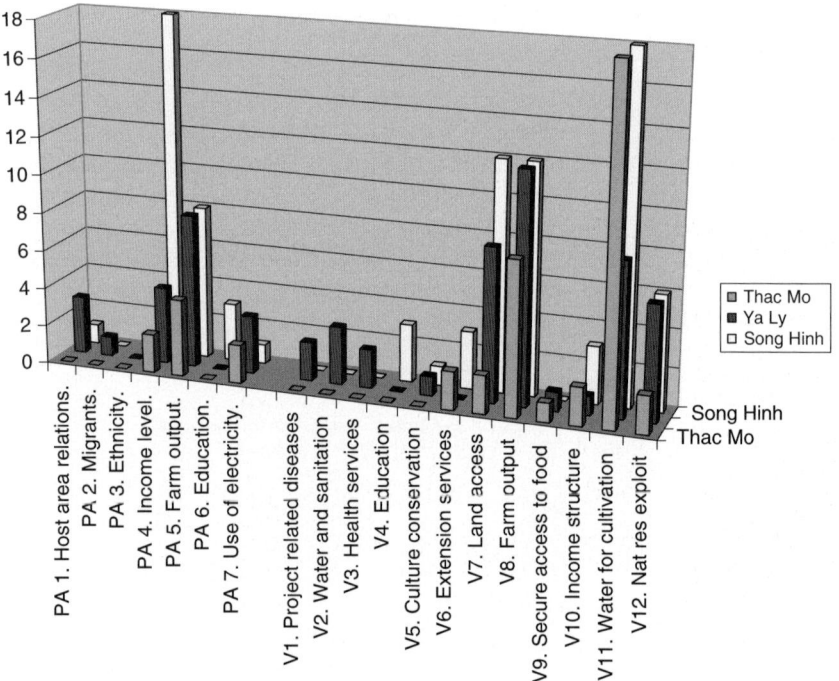

Fig. 4.4 The production/economic profiles in Thac Mo, Ya Ly and Song Hinh areas

at project area level is equally sensitive. The potentially serious issues experienced at village level are the in-flux of migrants, just like the case for the project area, together with farm output for marketing and land access.

Similar figures to those for Song Hinh can be drawn for Thac Mo and Ya Ly. Fig. 4.5 shows a summary where the differences between the three hydropower projects can be seen. In economic/production terms the situation in Song Hinh is more vulnerable than for the others.

These summary results are significant for how hydropower projects relate to the demands for sustainable development. Even if the purpose with the cited project has been to develop simple cost-effective ways to signal when extra in-depth studies are motivated, high scores such as for Song Hinh also indicate vulnerability and non-sustainable situations. The reporting is visual and transparent. It is well suited for consensus building around priority measures, including costing. The approach provides significant features for each of the project sites quickly.

The sensitive issues that come up are different in each case. For example, the rapid screening quoted above included one of the Song Hinh resettlement villages, Village No 2. It is a well-organized one, established through a collective removal of inhabitants in one village into pre-arranged new houses in a new location. Outsiders have not been invited to join the resettlement village. All households have been provided with electricity. This means that the inhabitants have a special background and a privileged treatment in comparison with the spontaneous village resettlements. This kind of special context is needed to interpret the findings correctly. In comparison with a regular assessment the model is of limited value; its strength is speed and cost, not accuracy.

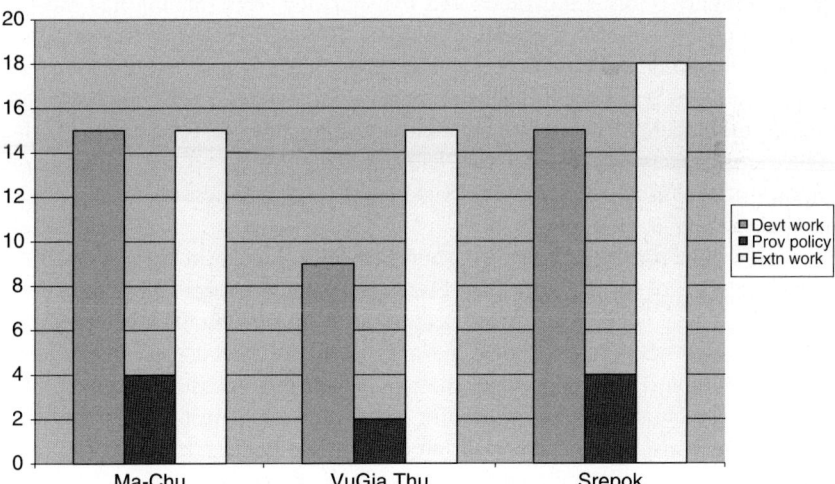

Fig. 4.5 Primary perceived competence for regional planning among stakeholders

4.1.3 Facilitating the Process

The account for experiences from three existing hydropower stations has illustrated an approach to Monitoring and Early Warning models. These models are applicable also at other hydropower stations. But they can be seen as the dynamic tools in the hands of active stakeholders. Designing the model itself contributes to sustainable development in the sense that a process approach is tried, with practical involvement by several stakeholder categories. It turns into a technique to build confidence, prevent conflict and come up with practical methods to identify problems before they have led to disaster. The only way to improve hydropower's reputation is by demonstrating a shift towards sustainable development. This shift has to relate to empirical evidence from stakeholder friendly implementation, rather than to policy documents.

Resettlees easily disassociate themselves with project planning until they experience real response on their issues. Similar problems almost emerged in the case of Nam Puoi village resettlement mentioned above. In this case the village was to be resettled on own initiative; not due to hydropower development, but because of an assumed land slide threat. In spite of an open dialogue with authorities, there was a perception by village stakeholders that they had little reason to get involved into what they felt would be a process above their heads anyway. Once a facilitation process got started, however, including consensus building across different stakeholder categories around both issue and solutions identification and prioritizing, mutual trust was built between administrations and villagers.

In the end all parties were appreciative and felt that the process proved very meaningful. The self-analysis showed that administration at various levels had trained thoroughly in decentralization and transparency. This training had mostly been carried out within each stakeholder category, though. The same was the case for villagers. Lacking was confidence in interaction, so that the fall-back into a traditional centralized approach to decision-making had occurred. With professional facilitation the stakeholders became forced to understand the different views of others. Once this process of understanding had been established, an intense series of dialogues opened up so that villagers could participate and contribute in design, planning and implementation.

The case of the three hydropower station staff involvement in the plight of resettlees shows a process where preconceived notions and barriers were removed. The diffuse and abstract picture of resettlement problems among the local staff was quickly sharpened when they got the opportunity to identify each issue and address its possible solution. This obvious awareness raising shifted externalized issues into the internal problem domain. This learning process is the main point for the current study, which also observes how different the situations can be at hydropower stations also within the same country and basically in the same part of it.

The next case study follows up on this regional level. River basin stakeholders discuss pros and cons with hydropower. The study focuses on the composition of the stakeholders (by category) and on their influence in the decision-making process.

4.2 Stakeholding Model (2): Regional Stakeholder Decision-making

The involvement by stakeholders is underscored in most development projects today. The gradually increased involvement over time has been noted for hydropower projects (Chap. 2). The current section addresses another aspect; who the stakeholders actually are. A key concern in this study is how influential stakeholders are in decision-making for regional or rural development. Both issues, who they are and their influence on project design, lead on to recruitment of participants; whether they have "a constituency" as representing one group in society. This can be a critical issue and has already been touched upon earlier.

The influence by stakeholders – whether participation is routine prescription or a working approach – is a key factor for sustainable development. A simple model has been developed below, based on the NHP Study. It illustrates real and perceived influential power of the stakeholders.

4.2.1 Stakeholders' Recruitment in the NHP Study

The participation in the various types of stakeholder workshops in the NHP Study was differently carried out. There were in all four different levels of stakeholding, designed as follows:

- *National* stakeholder meetings in Hanoi reporting to administrative stakeholders. Focus was on presentation and discussions on the various reports from the consultants.
- *River Basin* stakeholder workshops were held in each of the three river basins on information and knowledge. Workshops were held in a late phase of the project in order to also have the opportunity to assess findings.
- *Province* stakeholder workshops were held in all directly-affected provinces with participation both from directly-affected districts and communes. Focus was on data validation and on information. Workshops were held early in the project life in order also to sensitize and prepare for upcoming field work and village workshops. Participants were solely from the potentially affected districts and communes.
- *Village* stakeholder communications were carried out through participatory methods with focus on interactive data and information generation.

Attention is given here to the River Basin Workshops since these had the broadest rural development mandate. These workshops were held in the river basins of Ma-Chu, Vu Gia-Thu Bon, and Sre Pok. Regional issues, both upstream and downstream of each potential site were addressed at these river basin meetings. The consultations targeted all stakeholder categories, including relevant administrative levels. A detailed analysis of stakeholders' experience, potential influence and organizational involvement is given in Sect. 4.2.2. A comparison with their de facto involvement in decision-making is given in Sect. 4.2.3.

The stakeholder involvement included building consensus on key beneficial and detrimental consequences. Among the commitments was awareness raising through reporting back on project findings to the river basin inhabitants. This approach to build consensus around issues and solutions, potential and drawbacks, was invariably applied in all NHP Study stakeholder consultations. For the river basin case, differences in how inhabitants of the basins relate to hydropower project implementation were identified, their importance for stakeholders' livelihoods and life quality were ranked, and mitigation measures were identified, when possible. Also positive effects were addressed in a similar sequence: An inventory, a ranking according to significance and a discussion on integration into other development efforts. The river basin consultations had an expressed focus on regional development. This workshop series formed the consultations with stakeholders most closely geared towards decision-making.

Among the achieved results in the NHP Study was informed consensus building, where stakeholders contribute each their knowledge so that new revised consensus can be built as far as possible. Stakeholders dealt with benefits and drawbacks for regional development from a hydropower project Technical experts participated in district and river basin workshops, not primarily to give their views but to listen and clarify. Special attention was paid to resettlement and to the quality of the project's broader assessment of environmental and social consequences.

The workshops were preceded by stakeholder analyses, so that all potential categories according to the UNCED (1992) convention were identified. Not all posts were filled though. In reality it was EVN (Electricity of Vietnam) appointing the participants in all workshops except at village level. This meant a risk that stakeholders would be individuals elected on institutional rather than individual merits.

The stakeholder assessment included all workshop series of the project and the criteria for recruitment, as shown below in Table 4.2:

One side ambition in the NHP Study, along with decision-making and information processes, was to establish a Stakeholder Participation Model, SPM. This should account for stakeholders' constituencies; what interests they felt they represented. With this model in place, the participating stakeholders could be evaluated both in terms of what interests they represent and their impact on decision-making. A summary will be given below. The information sought about the participating stakeholders was three-fold:

- Past development project experience
- Degree of power in civil society
- Current involvement in the implementation of rural and regional development work.

These three factors combine into stakeholders' influence profiles. Even if the original design with open recruitment was not fully applied in the NHP Study, the interaction across stakeholder categories brought forth their specific issues. The following

Table 4.2 Stakeholder institutional background at planning workshops (national, river basin, district and village levels)

Stakeholder Category	National Workshops	River Basin Workshops	Province Workshops	Village Group Workshops
Women	Women's Union	Women's Union	Women's Union	Direct participation
Indigenous people	Political representatives	Political representatives, CEMMA[4]	Political representatives, CEMMA	Direct participation
Local authorities	National line ministries	Province level representation	Province and District authorities and commune People's Committees	People's Committee
Employees, unions	–	Farmers' Association	Farmers' Association	–
Business	Repr through line ministries	Representation	Female SMEs	–
Technology, science	Professionals from EVN	–	–	–
Land users	–	Farmers' Association	Farmers' Association	Direct participation

two sections (Sects. 4.2.2 and 4.2.3) give first stakeholders' perceptions of their influence, and then, by applying the SPM model, the realism in these self-images.

4.2.2 Stakeholders' Experience, Influence Potential and Organizational Involvement

This stakeholders' experience, their potential influence on decisions and how they are involved in development related organizations are summarized below. Findings about stakeholder experience, power and development involvement are also given. These results are then used in Sect. 4.2.3 for a formal analysis of stakeholder participation in decision-making.

4.2.2.1 Experience

Experience levels from development work, provincial policy, and extension activities are shown in Fig. 4.5 below (the y-axis shows the number of stakeholders).

---

[4] Committee for Ethnic Minorities and Mountainous Areas (Province and district level representation)

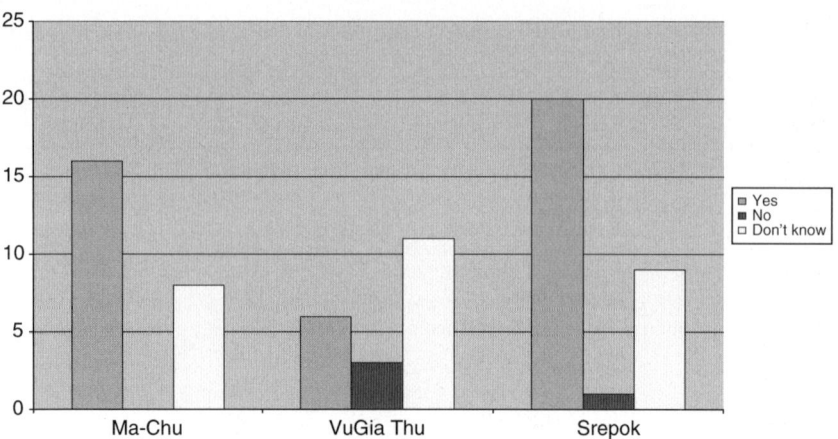

Fig. 4.6 Experienced capacity by stakeholders to influence decision-making *(Number of Persons)*

The pattern is quite uniform for the different river basins, showing stakeholders' practical experience from extension and development work, but limited involvement in policy formation or implementation.

The perceptions by the stakeholders about how much they can influence regional planning is given in Fig. 4.6: Each participant was asked about her/his experience, and how they considered they could influence provincial decision-making. This was a straight-forward question to the participants; yes, no or uncertain about whether their opinions have any influence.

The river basins showed different results. For the case of Ma-Chu, participants were fairly convinced that they had an influence. A similar view was held in Sre Pok, while Vu Gia-Thu Bon exhibited the opposite profile; most felt they could not influence decision-making, or at least doubted they could.

The professional background and the degree of involvement in regional planning are represented in Fig. 4.7

The impression is that the set of stakeholders who have taken part in the hydropower planning at regional level belong to a category of persons already deeply involved in government administrations.

4.2.2.2 Power

The River Basin Workshops focused on regional development. All 87 participants were interviewed in a survey in order to document their social, political and professional backgrounds. Since they had been hand-picked on the basis of competence, the selected group represents much of the intellectual potential in the three river basins, apart from representing the various stakeholder categories.

Fig. 4.8 suggests how socially and politically well established the stakeholders were by calculating the number of persons among the participants that each

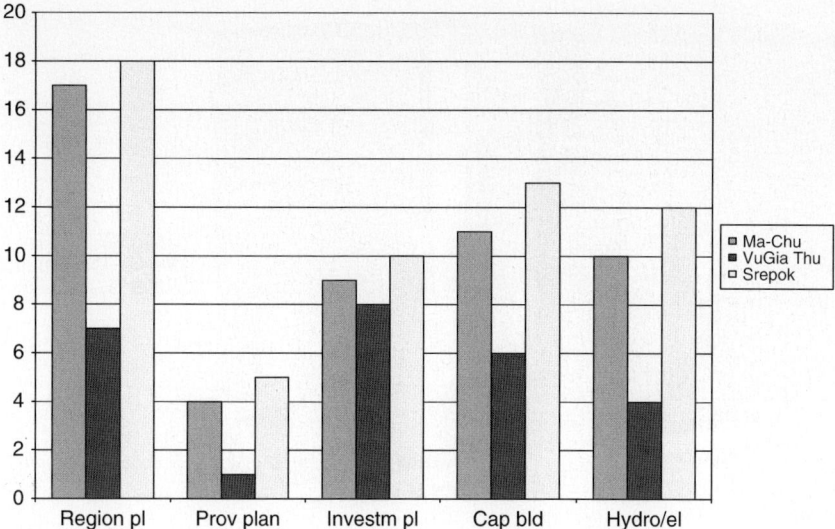

Fig. 4.7 Professional background of stakeholders at river basin workshops

stakeholder would know prior to the respective workshop. The x-axis shows the number of contacts and the y-axis the number of stakeholders. While naturally a number of relations already existed, it is noteworthy that participants did not come from one network only but were recruited from different social categories.

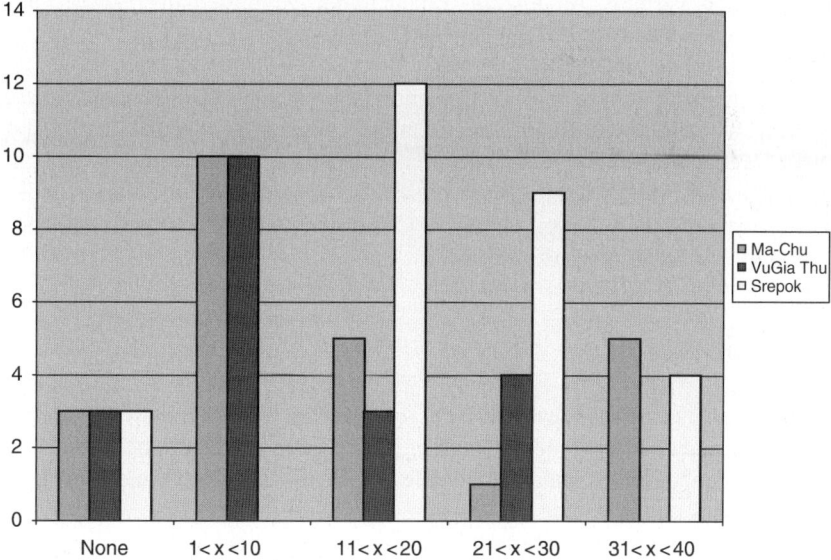

Fig. 4.8 Stakeholders' connections; networks and affiliations

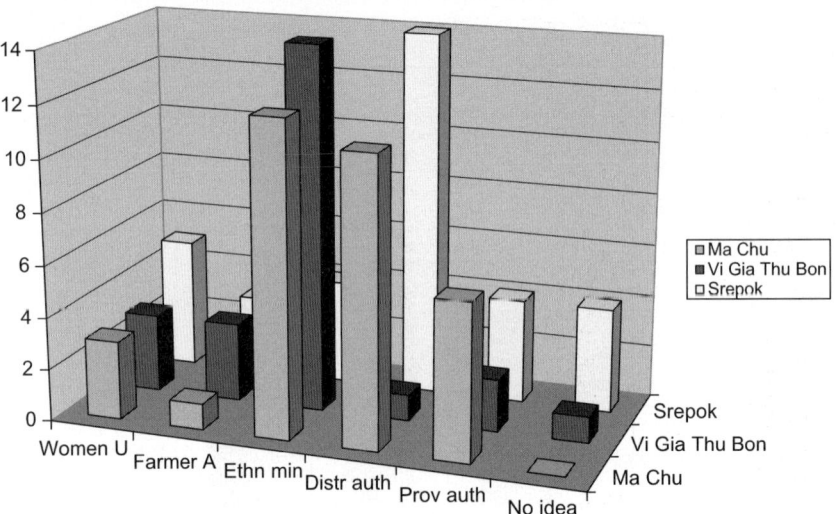

Fig. 4.9 Stakeholders' institutional affiliation

With the exception of Sre Pok River Basin, a relatively small number of contacts between the participants existed prior to the workshop. This suggests that the workshops succeeded in achieving broad stakeholder participation. Following up on the issue of representation, that is the affiliation of the stakeholders in key institutional or cultural terms, Fig. 4.9 below shows the width in how stakeholders themselves defined their institutional affiliation. The categories coming out were: Women's Union, Farmers' Association, Ethnic minority representative, District authorities, Provincial authorities, and No clear affiliation.

The Figure shows that ethnic minorities were very well represented, as should be expected with the demographic composition in the uplands. The least outspoken in this respect was Sre Pok River Basin. Another observation was that district and province level administration tend to be over-represented. The implication, though, is that local decision-making power is present in the stakeholding. This includes many members of DPCs (District People's Committee) and PPCs (Provincial People's Committee) as will be shown below.

4.2.2.3 Involvement

From the above can be seen that relatively few stakeholders felt they were directly involved in province level strategic planning, while many seemed to be engaged in extension or similar policy implementation levels. On the other hand a clear majority insisted that they can influence decisions in connection with regional or sub-basin planning (but the number of uncertainties in this respect is also high).

Going more local, to the combined district and province level involvement, gives the picture in Fig. 4.10

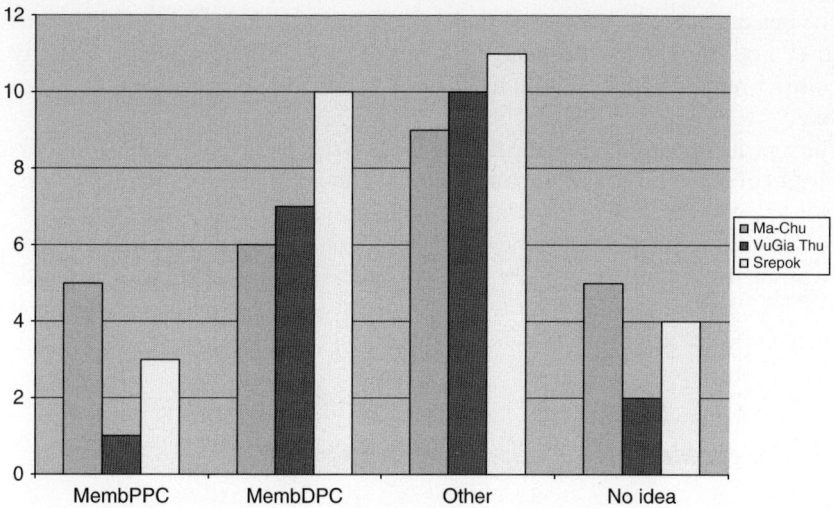

Fig. 4.10 Stakeholders' involvement in the operation of district and province peoples committees

There are two possible conclusions from the patterns shown. Stakeholders have good experience in government legislation and are well suited for dealing with the complexity of regional planning; notably balancing a region's growth with considerations for poverty reduction. Or, an interpretation may be that the participants are old foxes, going to many workshops, knowing the formalities but being less interested in the outcome of a planning process. It shows that they are very experienced in regional development (Fig. 4.11).

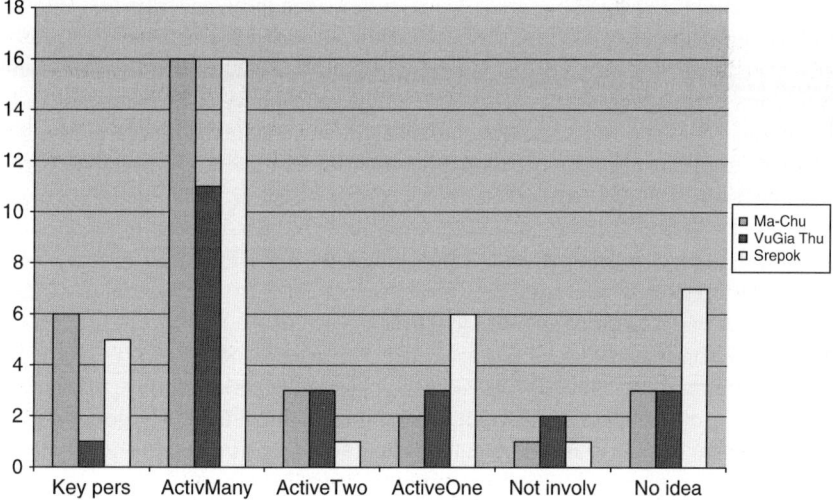

Fig. 4.11 Stakeholders' experience of regional development

The outcome may be connected to the participating stakeholders' affiliation as given in Fig. 4.9 above. Many of them identify themselves as ethnic minority and district representatives with limited involvement or role in regional decision-making.

The stakeholder profile from the figures above is that participants come from various walks of society. They do not form a closed group of administration supporters. They have good capacity to involve in forming information and taking part in decisions towards regional development with hydropower. Next section will compare this with the real influence they are able to exercise. Each stakeholder's own estimate of his/her position and power is documented and then compared with their real position, influence and decision-making power.

4.2.3 Stakeholders' Real Capacity and Power

Scrutinizing the involved stakeholders can continue into further detail for the various stakeholder categories (women, farmers, ethnic minorities, local decision-makers, regional decision-makers and business) by looking at competence, professionalism, organizational involvement, and experience from development projects. All this has been done in Phase 3 of Stage 2 of the NHP Study. A survey was carried out with all stakeholders as follows.

1. Experience. Four fields were listed, and the number of confirmative responses made up a score measuring experience. The four were experience from development work; position to influence decision-making about province policy; involvement in extension work; and other.
2. Decision-making. The decision-making capacity was expressed through three attitude questions: What is your professional profile (can influence: regional and/or provincial policy; province plans; financing/investment plans at province or district levels; plans for capacity development in development projects; hydropower development or rural electrification. The number of hits make the score for this question)? What is your organizational involvement (member of several boards or committees score 5; member of one board or committee score 4; attend board or committee meetings regularly score 3; attend board or committee meetings occasionally score 2; have no influence score 1)? What is your influence on decision-making (member of PPC score 5; member of DPC score 4; representative of a mass organization score 3; other score 1). The average value of the three scores was treated as an attitude score.
3. Experience. One question was asked: What is your experience of rural or regional development projects (key person score 5; active in several rural or regional development projects score 4; active in two rural or regional development projects score 3; active in one rural or regional development project score 2; not involved in any rural or regional development project score 1)?

Table 4.3 The power profiles of different stakeholder categories

Stakeholder category	Degree of experience	Potential power and capacity	Current involvem in devt projects	Sum
Women's Union	1,8	1,8	1,5	5,1
Farmers Association	1,8	2,1	2,4	6,3
Ethnic minority representative	2,5	2,3	3,0	7,8
District level official	2,6	2,9	2,6	8,1
Province level official	2,4	3,6	2,6	8,1

With these three features (perceived capacity, real capacity, and own experience) applied to each stakeholder category, a varied picture emerges. The scores per stakeholder category are given in Table 4.3. They are also presented in Fig. 4.12

The analysis of the different stakeholder categories shows a number of points where they differ:

- The perceived competence is by far highest among Province Authorities, followed by District Authorities. Lowest levels of self-esteem are found among Ethnic Minorities, Women and Farmers

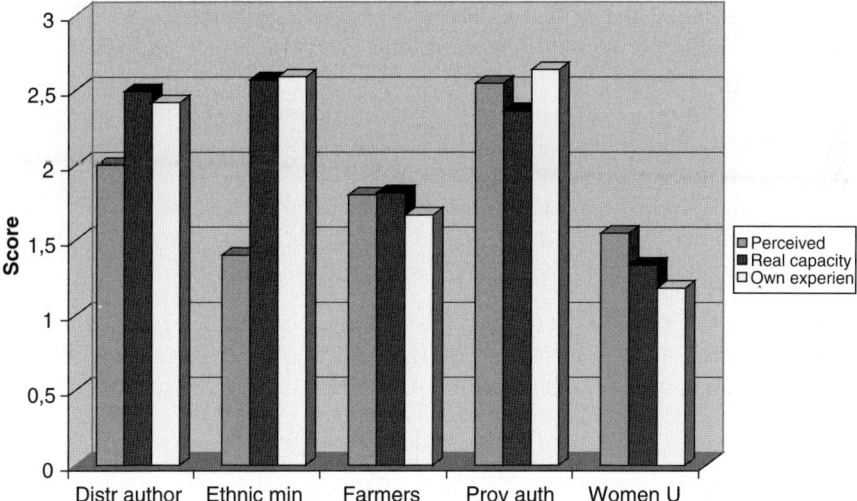

Fig. 4.12 Perceived and real capacity plus own experience among stakeholders of various categories

- The real capacity to make decisions with development or regional implications is different from the self-picture. Ethnic Minorities both take far more decisions than what their perceptions suggest. This latter performance is linked to the fact that many of the Ethnic Minority representatives are civil servants at district levels
- District Authorities and Ethnic Minority representatives have a modest view on their capacity. Women and Province Authorities appreciate their own competence relatively high in comparison with their performance
- Women and farmers are the two categories of stakeholders who are clearly less involved in real decision-making (for the case of women they believe that they are involved more than they are in reality)
- Farmers are consistent; they know their competence level and they are involved in decision-making up to this level
- In all instances the individual experience is approximately on level with the contributions people make to decision-making. It is considerably lower than what would be required for Women, and to some extent for Farmers. Provincial Authorities have more capacity than what is asked for.

The principles applied for how the stakeholder power analysis can be carried out are relevant for addressing sustainable development issues within hydropower planning. The issues just outlined for the river basin case (one out of the four stakeholder participation situations) are weighed and combined for each potential hydropower project. A weight can later be ascribed each stakeholder category, so that specific profiles can be represented graphically. The following characteristics for the stakeholders have been included:

- The degree of experience is assessed on an ordinal magnitude scale 1–5 for all projects (for all four identified areas of influence; river basin, national, district and village/commune)
- The potential power and capacity is assessed on an ordinal magnitude scale 1–5 for all projects (for all four identified areas of influence)
- The current involvement in development projects is assessed on an ordinal magnitude scale 1–5 for all projects (for all four identified areas of influence)
- The relative importance of the three stakeholder parameters above is weighted with the significance of prime stakeholder categories involved

The procedure can be repeated for river basin (region), catchment, project site and downstream areas. Thus each potential site is provided four matrixes showing quantitative (on ordinal scales) representations of experience, potential and involvement respectively.

With the stakeholder profiles thus established for each potential project site, a formal assessment is made of the stakeholder participation in the decision-making process relating to mitigation, should the specific hydropower plant project be implemented. Findings such as the generally good competence level imply a certain degree of success in mobilizing stakeholders. However, the recruited ones are generally found among formal organizations, already with a mandate to represent a political leadership. The result shows that ethnic minorities and women stakeholder categories are unclear; the self-image is that they have good influence while reality suggests that they do not. The general conclusion is that in particular women and ethnic minorities do not, in reality, exercise real political power. Such conclusions may be typical for the case of hydropower planning in Vietnam. The above study experiences are from Vietnam where the political organization makes the roles and mandates clear at all administrative levels with a hierarchical decision making tradition.

In other times, other locations, other countries the pattern may be different. But raising the issue is very important in that policy demands are to make local and regional interests meet, especially over economic growth and poverty reduction. For hydropower this whole process is firmly located under a national umbrella; that is the interest and need for energy production. Thus, mobilizing stakeholders into the planning process also brings the opportunity to revisit national strategies.

4.3 Stakeholding. From Prescription to Human Rights

The analysis of stakeholders' background and their self-images concerning their power position for providing good governance has demonstrated how both real and subjective potential to have an influence were addressed. But the NHP consultations were carried out not only with the specially selected number of workshop participants discussing development issues, but also with direct PAP in the villages, communes and districts to be directly affected. In villages, time allowed for a rapid sensitization, an identification of key issues (mostly relating to resettlement) and a documentation of peoples' considerations, worries and questions. The commune and district consultations also involved directly affected stakeholders, so that issues and solutions became limited to specific sits rather than concerned with regional development. For sustainable development a capacity building process is needed that could be continued at later project phases.

In retrospect, seeing several of the potential projects studied in NHP now being implemented, it can be concluded that this intention has not been fulfilled. There seems to be little continuity in stakeholder interaction in the concerned projects. On the contrary, the consultations appear as singular events, aiming at listing the assets to be compensated, rather than as steps in a long-term process. The question is then how all stakeholders, especially the directly affected people, can be drawn into implementation more systematically. The answer can be found in the kind of experience illustrated in Sect. 4.2.1. With the cooperation between directly affected people and hydropower staff with facilitation from consultants outlined there may follow a contribution to development of new approaches.

How stakeholder participation is implemented is strongly addressed and discussed in development projects today. It should fill a purpose, not be a goal in itself (cf. ADB 2004; quoted above). However, in hydropower planning the issue of involvement of project affected people in implementation can be brought further in hydropower projects. The multipurpose dimensions, beyond flood control, such as irrigation, water supply and sanitation, and fishing have tended to be treated as side issues. At least they have been given limited attention in the projects dealt with in this study. Under those circumstances the design of stakeholder consultations given in Sect. 4.2 can set a good example. But if the ambition is to have a multipurpose development impact beyond adjustments of roads and electricity lines, then a different style of approach is called for. The last chapter comes back to this, wondering how much actually has changed over a ten-year period: In any case the emphasis on multi-purpose requirements has increased as has the stakeholder presence.

In reality though, the rights and responsibilities of directly affected stakeholders are in the hands of implementing agencies. It seems that policy change has problems to penetrate a thinking based on predominantly technical views on how a hydropower project should perform; technically viable within an economic frame, and with a minimum of necessary adjustments for social and environmental complications. Even multipurpose contractual commitments, such as irrigation and flood control, seem negotiable in this tradition. For a project to be feasible not only in economic but also in social and environmental terms, all categories of stakeholders should be involved in consultation processes throughout the project planning, implementation and evaluation.

Chapter 5
Hydropower Implementation Performance for Sustainable Development

The implementation of Agenda 21 from Rio (1992) in the UN system, built on milestones since Stockholm (1972) and followed by the world summit in Johannesburg (2002), stresses long-term thinking where the interests of future generations are as important as the livelihoods of today's generations. Their key question transformed to the current study is: How are these policy and follow-ups having an impact on hydropower implementation? The case of hydropower planning is used to test the efficiency in Agenda 21 implementation. The choice of hydropower as "an indicator" is made because hydropower projects have far reaching consequences for key sustainable development issues, locally and regionally. Many institutions are involved, and the impacts from local conventions and policy change, consequently, have to come gradually. The study has provided examples on how changes in hydropower "protocol" have appeared.

The first concern of the study is whether there is an impact at all from new global policies – and the answer is affirmative. The second question is how new thinking shows up in everyday work life. The answer is given in the former chapters. Key social issues are raised or upgraded in their significance. The study has looked into a number of cases in order to outline trends for new knowledge formation. The third question is whether change is towards improved sustainable development, and if so, if the effect is lasting. That is the topic for the present, concluding chapter.

Conditions for information flow have improved tremendously even over the decade under the current study. For hydropower the following sections give brief but critical assessments of how information has been managed and not managed over the past decade. They look into the external perceptions of some of the studied projects. Attitudes are changing both inside and outside the hydropower sector. Furthermore, hydropower has received a symbolic side, representing a political stance in the environmental debates. What has been a technical engineering culture now needs to add the socio-cultural dimension. The chapter concludes on the contours of such a shift in a small but significant way. It notes how the experience from the cases in the study harmonizes with the needs for the world's sustainable development as expressed in the Millennium Goals.

5.1 Linkages Between New Information and Implementation

This study has concerned how additional social information is generated that is needed for proper hydropower planning to take place in accordance with the global conventions. Some modes of achieving suitably targeted information have been accounted for. These are aimed at reaching the new political goals in response to the environmental conventions. One question is if they have the capacity to generate sufficient reliable information. Another issue is if this information is properly made use of in project information. This issue is in focus for the current study.

The Swedish Sida policy on "Sustainable energy services for poverty reduction" (Sida 2005) is a donor interpretation of how political targets from global environmental conventions are expressed in development policy. The poor "... spend a disproportional amount of money and time on securing energy" (*ibid*: 2). It is particularly noted that reliance on poor quality fuels cause poor health. Poverty also leads to wood-gathering and deforestation. The policy conclusion is that access to appropriate energy services is a prerequisite for promoting pro-poor growth. Effective energy delivery systems are critical for achieving the Millennium Development Goals (MDG). Necessary elements in reforming a sustainable energy sector are changed ownership structures, transparency, and participation by relevant stakeholders. Institutional capacity building is underscored even to the point of stressing government interventions in energy markets due to "poverty and inequity, as well as environmental concerns..." (*ibid*: 4). Also all Sida support in the energy sector shall include environmental and social considerations.

These statements are the core of one donor's transformation of the international recommendations from the WCD into its policy on poverty reduction and the MDG. Swedish industry and government administration have agreed on a similar set of recommendations, "Swedish Waterhouse Guidelines". Sida is one of the partners. In short, these guidelines confirm the conclusions from the WCD and also emphasize that it is up to the project owner to make sure that laws, regulations and recommendations are followed.

All three initiatives aim at establishing good international practice with reference to hydropower performance in the era of sustainable development. The new demands for competence and knowledge include stakeholder involvement at all relevant levels in decision-making and data formation when social issues are concerned. This is stepping into new intellectual territory for main players around bringing hydropower implementation up to sustainable development standards. It shows for example in how they split on the issue of keeping the WCD as a set of recommendations and not make it legally binding.

This split is also reflected in the fact that there are now two levels of argumentation. Section 5.1.1 concerns the lessons learnt from the hydropower sector striving towards sustainable development. It concludes on how the new data and information systems can lead to improvements in implementation. Sect. 5.1.2 addresses the lobbyist environmental argumentation around hydropower as a destructive energy production system that should be stopped.

5.1.1 Building a Knowledge Base About Hydropower for Sustainable Development

The specific concern for the current study is how the new issues required by the global conventions have been interpreted in hydropower planning. The working hypothesis has been that response to the sustainability agenda is bound to show up in a search for new ways in response to critique and demands; new stakeholder roles, and new approaches (even methods). The study focus has therefore been on how new approaches are developed to access key social related information that is needed to build the required base for decision-making. The authorities, project owners, are assumed to follow suit with political intentions. The recommended information should be applied in project implementation. The environmental conventions are binding but broad in formulation. The hydropower related recommendations that have emerged during the studied period are in line with insights growing from the UNCED process.

The issues raised in previous chapters concerning hydropower project reform and sustainable development form a key in donors' and development banks' policy. This is particularly the case for poverty reduction and stakeholders' participation in the planning process as prerequisites for sustainable growth. The cases selected exhibit a gradual adaptation to new policy demands. New issues and new approaches appear. Judging from what the cases can tell, a delay in the adaptation to Agenda 21 for the implementation of hydropower plans is visible. An important question is if this is a temporary phenomenon in response to current political pressure. How flexible are the energy sector institutions to international sustainable development demands? Are energy agendas adaptable to changes?

The far-reaching effects from coupling World Commission on Dams (2000) with the global World Millennium Goals have been a widened agenda for hydropower planning. What was treated as project context 10–15 years ago, is now seen as a comprehensive set of key issues. The introduction to the current study lists them: Resettlement, downstream effects, poverty reduction, regional development, social safeguarding and market development, and concerns how these key issues have been addressed in a number of projects.

A general concern over the cases studied is to see how the (demanded) information generated is put into use in project implementation. This issue has been dealt with in earlier chapters, and leads to two further concerns:

- *Do the various actors have an interest and/or a capacity to incorporate the social issues into project design, planning and implementation? If so, what is their degree of involvement?* From "established hydropower developers" can sometimes be heard that the recommendations from global conventions are a constraint rather than an asset in practical work. The reason behind such thinking might be uncertainty about implications from the new requirements for an established working style. The capacity of involved actors to address and understand the consequences can be limited. In particular, the socio-cultural issues addressed in the current study call for an increased awareness when building professional confidence in all institutional levels of implementation. A number of criteria for good governance can be used to meet minimum requirements for socio-cultural attention. Their function can be to contribute to the BCR calculations that set the scene for technical and economic feasibility. What is also needed is a work style that automatically internalizes the sustainability thinking. This step, reaching beyond the check-list stage, is called a paradigmatic shift here. The general conclusion is that the trend in the studied projects is constructive towards such a shift, but that it takes long time.

- *Does safe-guarding only aim at minimizing negative effects, or can it be seen as a regional development level concern, where a hydropower project forms an integrated part of a development process?* In the ongoing development process hydropower planning needs to incorporate social and ecological environment into a design process for sustainability, meeting simultaneously the three basic requirements of economic, environmental and socially sustainable development. Balancing acts are necessary, for instance when policy says economic growth with due consideration of poverty reduction goals. Hydropower projects need to be further moved into such a process. There is a risk, out of uncertainty in a policy change situation, to rely on checklists and upholding minimum required level of protection for affected social groups. Such an observation gives an impression of short-term performance. This speaks rather against the anticipated paradigmatic change.

The two issues in italics are discussed in Sects. 5.3 and 5.4. The first one concerns a broadened outlook towards a more holistic and future-oriented thinking in implementation. Secondly, such new thinking has to incorporate not only what is project consequences today, but also the future, so that safe-guarding concerns not least future generations.

5.1.2 Disinformation and Criticism as Political Language

In the hearings in Windhoek on the Epupa project (see Chap. 2) there were comments from both the press and the dominating NGO, International Rivers Network, that the information provided by the consultant was irrelevant since large-scale hydropower projects by definition are environmentally destructive. The alternative renewable energy source of solar power was raised but proved to demand too large land areas. The off-shore gas fields of Namibia were pointed at as an alternative for that country. But little was mentioned about the Angola interests in the context of this potential border-river project.

A similar debate is emerging for Vietnam and downstream situations in Cambodia. One difference may be that the main alternative options are different; for Vietnam it is nuclear power. The same, or similar, positioning by the same, or closely related, international NGO appear in the issue of hydropower construction along the Sre Pok river on the Vietnamese upstream basin. Two of the river basins in the second stage of the NHP Study were international, concerning the neighbouring Cambodia and Laos. In the Cambodia case a special screening of downstream effects in the Sre Pok river basin was initiated. A Canadian NGO, Probe International, as well as a Norwegian journal, Development Today, did mistakenly see the study as an EIA and criticized it on those premises. This misunderstanding has been launched on the internet and not been corrected in spite of several reminders that the NHP Study is not to be seen as an EIA.

The reason for such tactics is political. Critical talk is the way to bring home a message to stop hydropower planning. Seeking correct information is in this process not the level of interaction. Instead, interaction is through statements and political clarifications. This goes for donor governments as well. The main decision-makers, the governments of the concerned countries, are not the main targets in this kind of debate. Instead, a pressure group pattern emerges. These are the groups with an agenda to stop hydropower planning through attacking the donors (who in their turn have an agenda of a more sustainable approach than the national governments and companies implementing the projects).

By sticking to biased information political pressure is established on the donors. On this level also keeping quiet and not responding is a message. Sweden's late response can be interpreted as ambivalence, which in turn signals that critics have an issue. The issue changes from scrutinizing whether international conventions are followed into implementing political ideology.

The role of the NGO community is strong, both for the sustainable development process and for a spill-over into hydropower planning projects. Ideologically, NGOs span a broad field. In terms of political influence they are highly significant, termed even to be one of the world's superpowers. The other side, the proponents of hydropower in the shape of industry, also has formed an interest grouping in International Hydropower Association (2004). That network provides policy statements in line with the WCD. The reference is a guide to operators and managers of hydropower projects. It contains the principal thinking behind sustainability (balancing environmental, social and economic aspects) and commits members to a policy statement: "Eradicating poverty, changing unsustainable patterns of production and consumption, and protecting and managing the natural resource base underpinning economic and social development are overarching objectives of, and essential requirements for sustainable development." (ibid: 3). The guidelines basically adhere to the core of the WCD but disagree with some of the detailed recommendations.

With reference to social issues connected to hydropower development the guide states that hydropower schemes "have the ability to significantly reduce poverty and enhance quality of life in the communities they serve" (ibid: 17). It follows up with recommendations on managing social impacts, the outcomes that should be expected, and strategies for reaching the goals. So here is the other side of the rhetoric, carrying a message of rationality: Through a heavily structured and objectified text the message comes that hydropower planning is under control, potential unexpected difficulties can be handled, and there is no reason for worry over the implementation.

Both inputs, nature conservationists and rationalists, may contribute to a polarization that is not conducive for an open dialogue between stakeholders. The current study on how sustainable development conventions policy is transformed into action, using hydropower as the example, can note that there are other information systems pro and con hydropower development that are in operation. This is an obstacle that can be noted even though it will not be penetrated further in the current study. The following section goes instead into the ground level of obstacles by drawing on the experiences from previous chapters and from the actual implementation of projects.

5.2 Complications in Linking Implementation with Policy

The recorded communication difficulties between various actors give rise to a concern over how policy links with implementation; a main theme for the current study. The following observations on linkages have emerged out of the study so far, through the use of hydropower as an indicator of impact from global conventions into project planning.

- The stakeholder involvement in the NHP Study (see Chap. 4) increased drastically during the project implementation. In Stage 1 (1995–1998) their participation was just hardly accepted by the Client. Stage 2 (2004–2006) opened up for a more systematic stakeholder participation. The Consultant's proposal to extend this participation beyond the contracted engagement was accepted by the Client. *Conclusion*: There is a positive trend involving stakeholders. This involvement is not an end in itself but needs careful targeting. It comes close to involving affected local communities, invariably formed by ethnic minorities, but stops short of linking up with implementation.
- The Consultant's proposal in the NHP Study, also written into the contract, that stakeholder recruitment should take place independent from the Client (an election-style approach was proposed), was in the end rejected by the Client. The selection of NGOs allowed to participate was very tight. *Conclusion*: The client can not afford independent stakeholder participation in large-scale hydropower projects.
- The prime donor of the NHP Study, NORAD, emphasized two key problem domains: resettlement along with other migration issues, and downstream effects. NORAD also emphasized the importance of the data base built by the project; that it is kept updated and utilized for upcoming individual projects. In a discussion with the two lead advisory persons to EVN (the Client), their interpretation is hard to follow. Both issues have severe cost implications if mitigation alternatives are long-term (compensation and regulation). *Conclusion*: The issues of downstream effects and resettlement break away from sustainability indicators into cost and financial concerns.

- The first opportunity to apply the findings from the NHP Study for EVN was the ADB financed Song Bung 4 feasibility study (see Sect. 5.3.1). One positive impact was that the consultant to ADB made extensive use of the primary data gathered by the NHP Study when writing the feasibility study. However, the desk officer and ToR author confirmed that she had not even read the NHP Study report in her preparations of the Song Bung 4 project. *Conclusion*: donors and client need to improve coordination between projects.

- Sida (Swedish co-donor of the NHP Study) appointed a special evaluation team from its Environmental Help Desk at the Swedish University of Agricultural Sciences (SUAS).[1] The process of analysis by this team began with a ill-informed desk study assessment and continued with a more positive interpretation of the NHP Study as the team members increased their insight into the project. However, the net result, to give positive assessments of both the environmental and the social components did not reach deep enough into several parts of the project. *Conclusion*: The complexity of issues is difficult to overview in a short period of time, consequently enhancing a technical dominance.

- Both the Isiolo (Kenya) and Lesotho studies in Chap. 3 touch upon the significance of SMEs for regional development and poverty reduction. *Conclusion*: It is in line with the global conventions to link hydropower development with regional efforts to reduce poverty through economic growth. Providing rural electrification is then a suitable form for mitigation, such as the experiences mentioned in Chap. 3 from Nepal may illustrate.

- The trend over a ten-year period is from check-list data collection towards stakeholder identified issues. *Conclusion*: Impact assessments need to be specific by relating to each individual situation. Stakeholder involvement is crucial, and must aim at an interaction process with various set-ups of the stakeholders.

- The experiments with relying on targeted information have succeeded at pre-feasibility level assessments (and more in-depth for the specific issues raised). The experiences of modeling in early warnings this way have also been successful. *Conclusion*: If used with caution and linked with further studies, the rapid screening techniques into new issues can offer effective methods for consideration of emerging issues within hydropower planning.

- A prime intention with this study has been to use the introduction/acceptance of new methods as an indicator of the project owner's response to the global conventions. *Conclusion*: The correlation between the methods, such as upgraded stakeholder involvement and more penetration into long-term resettlement and downstream situations, and global conventions/development policy is good. However, a whole paradigmatic shift is needed in project leadership. It is not viable to expect technically trained persons to lead social assessments without indepth insight into needed approaches and methods.

- One particularly significant problem in this respect is the gap between data formation and stakeholder involvement. The study shows good progress for data

---

[1] SLU, Sida helpdesk: Monitoring of environmental and social aspects in the national hydropower plan (NHP) study in Vietnam (Wallentinus et al. 2005).

formation. Information has become more targeted and relevant for the major social issues in hydropower development. However, the effort to establish stakeholder processes in other aspects of sustainable development has not gained much foot-hold in hydropower planning. Feedback of useful data to the benefit of the Poor as a target group, such as development plans and infrastructure investment, has only been applied in a limited way. If data generated for project implementation had also been available for other regional projects, much needed synergies could be achieved for the sake of coordination and integration. *Conclusion*: Efforts to establish a data bank, as in the case of the NHP Study, must be encouraged, and the information used for building up an insight of both problems and potential within hydropower development.

- The involvement of stakeholders in project planning, not only data formation, is a complicated matter since so many interests are represented. As just mentioned, there is nowadays a general insight beyond hydropower into the usefulness of stakeholder interaction among project management. There is, however, also a restraint to provide insight into the very same management. *Conclusion*: Stakeholder involvement must reach decision-making, so that ownership of important project aspects, like regional development, can be established. Such a process requires considerable learning process of the various parties concerned. Affected stakeholders should need a mandate and to represent an interest category, maybe even a kind of "constituency". Since a consequence is that project costs will increase with upgraded involvement, proper policy backing is required.

- There is today widespread international attention to large-scale hydropower projects. This must be related to sustainable development in such a way that a holistic view on water resources integration can form the hydropower context. *Conclusion*: A formula for making projects accountable also to the global conventions could be worked out on the basis of the WCD. This naturally includes considering multi-purpose alternative water use.

- The integration of water resources management is commonly accepted as a stepping stone. Hydropower reservoirs should be multipurpose in design. In implementation though this is less clear. IWRM calls for upgraded sustainable development. *Conclusion*: Sustainable development indicators accounting for environmental, social and economic (also beyond B/C ratios) should be developed for the purpose of rapid screening of potential project options. The NHP Study provides a good start for methodology on this.

- The multi-purpose issue for directly affected people leads to the broader benefit-sharing issue seen in a regional development perspective. *Conclusion*: Could benefit sharing be further developed? It is a big issue that will emerge more and more the coming years. Downstream situations take the lead, as accounted for in Chap. 4.

The list of obstacles in correlating hydropower planning with its implementation could be made longer. Yet the concerns above serve to indicate the fact that there is a serious dissonance between design intentions and implementation reality. This is, of course, to be expected since the topic under study is the paradigm shift imposed on the hydropower production sector. The WCD is the prime forum for making the different views on hydropower development meet. It is seen by some stakeholders as the embryo to an international compulsory policy. In the committee mentioned above, formed by prime stakeholders in hydropower production in Sweden, national recommendations based on the WCD have been worked out for future domestic production (Swedish Water House 2005). Similar efforts come from other countries. But worries obviously remain over how serious such recommendations become as long as they are not binding. For instance, do they apply to consultancy firms and industry operating within development cooperation or to hydropower production in regional cooperation.

Even if national policy issues are beyond the scope of the current study, examples can be suitable for understanding the impact of sustainable development targets. What has occurred so far, however, seems to be a limited adaptation to global constraints based on agreed conventions. The trend in implementation remains maintaining a conservative benefit/cost ratio indicator as the prime issue when assessing the viability of a potential project rather than drawing on a full set of sustainability indicators. Critics from the South often state that it is appropriate for countries in the North, where much of the hydropower potential is already developed, to follow the WCD recommendations; but developing countries today would not have an equal opportunity if they were to follow the WCD. They naturally observe that while Western Europe has developed 85 % of its hydropower potential, Africa has done so with 5 % only.

This comment reflects a current process in which the issue of dams and development is about to be filed into the classic North/South controversy. Rich countries are expected to contribute to the financing of solutions based on sustainable development approach and not on commercial benefit/cost thinking. If the day comes when this pressure will result in rigid implementation of the WCD recommendations, it will indeed be a political necessity to clearly understand, and consider, the socio-economic issues. In that situation it is not sufficient to hide the social concerns inside environmental assessments as a factor where negative impacts should be minimized. Instead, the interconnected social and environmental matters need to be lifted to the fore, and important questions should concern the most suitable management of water resources. This is then not only in terms of water use but also in terms of regional and national development where various infrastructure inputs can be combined. A further question is how national policies could be adjusted to the global conventions, and what kind of international safeguarding mechanisms should be developed in order to influence national policies and, consequently, implementing agencies towards a more holistic, and environmentally and socially sustainable hydropower implementation.

5.3 A New Holistic Perspective in Technical Culture

Can the hydropower development sector be more responsive to the Millennium Goals? Obviously, one factor is that costs will increase if implementation shall follow a range of recommended international principles. But it can be argued that these will be covered involuntarily by the affected populations unless included in project budget. In the projects accounted for in this study there has been no debate over a possible level of response to the WCD recommendations that would be socially motivated. Instead, when WCD has appeared among implementing agencies it has been primarily in connection with new needed design for mitigation and stakeholder involvement. The recommendations have been treated like a *smorgasbord* offering appetizing measures; stakeholder analysis, conflict resolution as seen fit in a specific project planning. There has been no effort to transform the recommendations fully into design and implementation.

A broad development has taken place in the hydropower sector towards a higher level of social responsibility. This has already been noted in the study. The improvement trend seems to be stalling, possibly due to differences in agendas for the various actors involved. Missing out on negotiating these differences and resolving them leads to a situation of ambivalence. The scene becomes set for alliances between actors with a mutual interest but based on different views.

5.3.1 Ambivalence in Understandings of Policy Intention Open for Interpretations

One of the reasons to the great variations in how well-meaning experts transform policy intentions into practice, is their leadership. This in turn suggests that perspective, not only individuals, might show differences in interpretations. In such situations of ambivalence, key issues like downstream effects, fishery, village consultation process, forestry, livelihoods and life quality usually have not been agreed upon at the project start-up. This was the case, for example, in the Song Bung 4 feasibility study. Even more complicated for implementation are changes in issue foci in Song Bung 4 and also in the 2RRBSP, Part 2, project. Both exhibit major changes in ToR in the midst of implementation; combining ethnic minority and resettlement studies, and adding gender considerations respectively. The Song Bung 4 project was initiated before the client even was clear over its own intentions. Not only the re-writing of ToR but also changed deadlines and changed reporting structure gave rise to frustration both among affected people (stakeholder meetings called for three times and called off in the last minute) and consultants (changed work plans impacted work with other projects). Such changes were not caused by any decision-making processes with stakeholders but rather by indecisive leadership. The result can be inconsistencies and information gaps, and in the end even deviation from both policy and implementation design.

5.3.2 The Lack of Checks and Balances in Implementation

Policy and intentions for implementation have responded to the global conventions over the ten years under study. The implementation cultures developed over long time sometimes stand against the intended changes from UNCED and WSSD, including the WCD. Especially the major stakeholders in hydropower planning are careful to make their positions clear, as stressed by the International Hydropower Association. Apart from industry there are different interest groups behind the WCD recommendations. These are equally important in the process of hydropower project planning, and perhaps also in implementation. There can very well be conflicts of interest. The WCD sought to bridge these, but many issues remain between key actors, such as:

- Governments
- Line ministries
- International consultant companies
- International NGOs
- Donors
- Development banks
- Pressure groups and mass media.

A few of the studied cases in this study have been much debated projects that can provide illustrations to the observation of conflicting interests of various actors. The differences between actors have provoked positionings by them, either openly or by implication. The following examples illustrate how these different players may act:

Governments have both broader policy implementation to look after, and the responsibility that laws and regulations are followed in hydropower planning and implementation. Their cross-cutting interest in poverty reduction is extremely high on the international agenda, and also reaches hydropower planning. There are good political reasons for this, beyond the desire to protect vulnerable populations: Poverty situations signal many global problems, since poverty creates suffering, leading to unrest and resentment. This fact has become a strategic issue in global sustainable development in that the consequences foster national and international terrorism through a frustration over chronic poverty. What is more, hydropower sites more often than not are located in remote mountain areas, they have long felt politically marginalized, being a playground for skirmishes and embryonic civil wars, and battlefields for conflicts between neighbouring countries.

The governments have to address several key cross-cutting issues in sustainable development. Poverty reduction is a very important example. It is addressed by international NGOs and consultants alike (cf. GPF and NGO in internet). A recent assessment to the EU done by an international NGO of the increased poverty reduction aid claims that very little of the donor increase actually reach the Poor. Instead the layers between the donor (EU) countries and the targeted poor populations absorb the resources (Source Viet Nam News, May 15, 2007). In this respect governments' drive towards increased sustainable development seems limited.

Line ministries are carriers of a stringent company culture. Visiting the electricity companies of Vietnam, Namibia, Angola, Lesotho, Costa Rica or Nicaragua, the countries appearing in the current study, gives a feeling of a *déja-vu* of the same company culture; reflecting an interest in technical and economic efficiency. Environmental and social considerations are in such quarters sometimes still treated as externalities that need to be addressed simply because they nowadays are conditional to financing. Checks and balances refer strongly to the "own" administrative sector.

International consultant companies have the benefit to follow a work plan based on already detailed Terms of References. The ToR, rather than reality, are followed if conflicting interests occur, such as through the appearance of unexpected circumstances. Since Consultants work on a commercial basis, all major changes must be accompanied with negotiations about changes in the ToR. The studied projects have some instances of initiatives from Consultants to upgrade or supplement their original ToR, but normally a consultant follows them rigorously, not least since they are legally binding. Maintaining checks and balances then has a contract rather than a reality as reference frame.

International NGOs have in several of the hydropower projects studied criticized the Consultant rather than the project owning Electricity Company. Both the NHP Study and the Epupa feasibility study offer examples. In Epupa, International Rivers Network, IRN, had requested a biased "independent report" with serious mistakes about the project design. In the SrePok follow-up of the NHP Study, Probe international, an NGO closely allied to IRN, followed the same style, providing an "independent" report with major mistakes. See also Sect. 5.1.2 above. A plausible interpretation is that desire to maintain a radical and systems critical profile over-rules a search for objective information. In consequence, proponents are reluctant to go into the various specific issues. If doing so they risk to end up criticizing governments. A hidden agenda for them could be to maintain the opportunity to operate in the countries involved, and so, perhaps unconsciously, refrain from addressing any root problems. In this respect the NGO performance can be characterized as lacking both in checks and balances. It adds to the "political party" performance with NGO's profile as proponents of a stand rather than being involved in an issue solution process.

Donors have a political constituency in their own countries. Desk officers relate to the development discourse "at home". These two factors show clearly in the cases of the NHP Study, again, and in the international rivers issue between Cambodia and Vietnam through Pok river. When a new government is installed, as for Sweden in the end of the study period, new policy thinking is revisited and assessed. This is felt in the donor end as a drastic change in policy, so that even single speeches by political key persons are listened to and interpreted, as a way to forecast policy change. Otherwise, project assessments are contracted outside the donor organization. In the case of Sida, this was contracted for the NHP Study to the Environment Help Desk at SUAS. The independence requirement led to, as in the cases related to the NGO case of the International Rivers Network, that the Consultant did not secure having mistakes in reporting corrected before publishing. The objectivity search did not open for checks and balances in facts reporting on management.

Development banks have their multinational policy principles, and national pol- icy might differ. This is the case for Vietnam. Like in the rest of South East Asia the main actors are the World Bank and the ADB. The differences in policy come on the agenda during project negotiations. The ways they are treated relate to that situa- tion. The discrepancies are not trashed out in search for a best solution. Instead they drift into negotiation rhetoric, where the bank side may set up internal minimum requirements for agreement. No stakeholders involve in that process.

Banks differ from the donors in that they operate for profit. They do so with a half-professional stance, shaped by the conditionalities formed by development community membership. This could also open for ambivalence; the actors can pos- sibly operate with low-level checks and balances, as hinted for ADB in connec- tion with the Song Bung 4 case. Here, it took the project's social section some two months to get off the ground simply because the bank team kept modifying the ToR of the recruited consultant, and even brought its internal debate over what should be done into the lap of the consultant. This kind of performance is not limited to ADB's hydropower projects, but occurred also, for instance, in the 2RRBSP Part II, where a gender issue was added late in the project.

Pressure groups: NGOs and mass media. Mass media often link very close with international NGOs. However, their mandate is different – while the NGOs would see as their task to raise opinion and attention, mass media should provide informa- tion. There are good examples from the studied cases, such as the four films about the Epupa project, of making information available to the public. The film produc- ers keep a distance from the project implementation by accounting for facts and dilemmas in the project process. In other situations in the studied projects NGOs and mass media come closer to each others' goals when looking for targeted infor- mation. They both have a power base in their information control/management. In the developing world the NGOs are turning into being the second world power (see Sect. 5.1.2), next to the US. In the North mass media are since long powerful. In both cases the accountability is to the general public but they also have the potential to form opinions. In this respect checks and balances may appear diffuse, especially in a short-term perspective: Criteria for accountability are derived from an audience's valuation of reporting. Its political stance is one factor just as if there were radical and reactionary hydropower projects. Attitudes and ideology issues are addressed in next sections.

5.3.3 The Culture of Sustainable Hydropower Development

The process of adaptation in the hydropower sector to recent global convention has not least sought the degree of inclusion of social issues in planning and im- plementation. The cases covering ten years suggest that this is a slow process. The difficulty more generally, for all water sub-sectors, is to achieve a lasting change in development paradigms for water resources development. This seems to be one reason behind a recent (2006) initiative by the UNESCO International Hydrology Program to establish NETWA, Network for Water Anthropology. NETWA connects socio-cultural studies with water engineering for obtaining the Millennium Development Goals. It is a new design to which the observations in the current study fit well. In order to be consequent with sustainable development; the various water sectors need to relate also to social (and environment) issues. IWRM has to integrate not only water sub-sectors but water resources management with all sides to sustainable development as well.

This is also the issue, specifically for the current study, with the hydropower sector. The gap in perspectives is indirectly confirmed for this sector through several recent observations, for instance by WWF (2005). This organization shows, in the wake of the WCD, how a gap remains between rhetoric in the form of upgrading social issues, and the implementation of these issues without prior thorough capacity building among the international experts.

An overview over how the selected projects in the current study relate to UNCED, WSSD and WCD has been given in the introductory chapter. Most projects can link up with these three relevant global agreements/recommendations in one way or another. There is a positive trend towards more attention paid by the consultants to the three global approaches (convention, recommendation and summit for agreement). Pressure has been built by donor organizations and by NGOs, or it comes from within consultancy firms. The WWF (2005) report asserts this positive trend in a global perspective with verification studies of a few current hydropower projects. But the report also claims that the progress during the first five years after WCD is only gradual and far from success:

> WWF welcomes these positive developments but also deplores that some dams are still built based on dubious economic arguments, without considering all alternatives, without transparent processes and without adequately addressing serious environmental and social impacts, as demonstrated by the six case studies in this report. (WWF 2005:14)

A concern to be addressed is how effective such reporting is. Publicity seems to be one of the few corrective measures available, assuming the actors listed in Sect. 5.3.2 look after their particular interests. The section indicated a concern with their accountability. Accountability can be seen in several ways, economic of course, but also democratic and ethical. Since there is no equivalent to a court of justice where appeals can be made of disinformation, flaws in policy application, or other deviations from a set of "rules of conduct" are difficult to address. The recent tendency towards good practice within hydropower planning could become a parallel to the efforts within the UNESCO system by industry. That would in turn put pressure on industry to perform with delivering documents that are less sweeping. When these developments occur the desired shift of WCD recommendations into implementation could become reality. The quote reflects what is the only hope – that the emerging process of change that follows on the environmental conventions will lead to the best practice possible towards sustainable development in the hydropower sector.

5.4 Hydropower and Environment Ideologies

Last section referred to some rigid positions in the emerging change in the hydropower sector. It concluded with a plea for a technical culture based on good practice. Included in this practice are the new demands of targeting poverty reduction, involving all stakeholder categories in projects, and transparency in both formulation, planning, design, implementation and evaluation stages. Through large-scale projects, such as hydropower, information (identification and data formation) and implementation of data jointly with stakeholders can be developed. The previous section also noted that there might be several political agendas in the case of hydropower development, which complicate studying the processes currently taking place. The fact that some of the lobbying against hydropower planning is based on a fundamentalist ideology that large-scale hydropower projects are bad and to be stopped leads to the issue of environmental ideology discourses of relevance for large-scale projects concerning sustainable development policy, which is the topic of the current section. It revisits some of the cases under study in order to illustrate with expressions of what boils down to environmental conservationist ideologies.

The environmental debate relating to the studied projects is fuelled by the global changes the world experiences, climatic as well as in economics, informatics and technology. Energy demands and energy needs are high and have to be met. Poverty reduction is one conclusion in the Millennium Development Goals, which could be seen as a nicer way of expressing the security interests of the North. The ecological and social effects of large dams appear in this context and are striking. The database for decisions to embark on hydropower therefore needs to be supplemented with an analysis if this really is the best practice to meet energy demands, or if there are alternative energy sources.

Furthermore, a basic question is also the rationale for decision-making: Is all competence found among stakeholders, high and low, focused so as to safeguard best decisions and best mitigation? There are obviously also other considerations to be made, relating e.g. to lifestyle and value systems.

In his critique of dogmatic thinking among environmental NGOs, Arne Kalland (1998) has coined the *super-whale* concept as representing the totemistic side of environmental rhetoric. The main argument from the whale hunting sector is that critics combine environmental issues relating to different situations (species in the whale case) into one simplistic argument. Like with whales (seals included): While there are many whale species living under different circumstances, there is mentally only one "super-whale" species in the minds of critics. Conservationists' ambition to protect expands beyond reason, threatening and destroying livelihood systems that actually are sustainable.

The purpose here is, of course, not to enter the whale issue. However, it illuminates a phenomenon apparent also within hydropower development. Both relate to the conditional sides to sustainable development – socially, ecologically and economically sound natural resource use. There is in other words a totemistic side to hydropower – a super-hydropower plant that can be seen as destructive to its environment and should be abolished.

The illustrations below concern tendencies towards the ideological ways of perceiving hydropower planning as intrinsically bad. The intention is not to make rigid political analyses but simply to present where differences in opinion may relate to the information management. The following three cases have been touched upon:

- Poverty reduction – ideological positioning with a hidden security agenda (a span of opinions from the UNESCO message that poverty is reduced to the NGO observation that those worst off get it even worse)
- Information availability (from the international lobby movement against the project in Angola and Namibia to refuse all information, to the pseudo-scientific assessment of the NHP Study by a Sida institutional consultant)
- Stakeholders' involvement (from the NGO preoccupation with local participation via authorities preoccupation with institutions to consultants preoccupation with technical and economic competence).

The varied views in these three examples demonstrate how vested interests may penetrate problem perceptions and block a debate on issues and solutions. The language spoken is the political language where goals are already set by ideology and information is filtered to support the goals. The relevance for the current study is the tendency to block open information flow through the appearance of "sector cultures". The ideologies influence how issues are perceived. The hydropower sector becomes a vivid example of obstacles to the growth of a sustainable development culture in its different actor quarters. Facts and information become branded as suitable or not, classified or not, and lacking or not.

5.4.1 Poverty Reduction – Budget Rhetoric or Solidarity?

A major rationale in all development aid cooperation today is that poverty must be reduced. This is a strategic goal, also with improved international relations in mind (see Sect. 5.2). Donors allocate funds, tax payers' money, that go into heavy institutional systems. The political message the governments give is that they show action and consequently hope for partners' response so that the goal is sufficiently achieved. They are accountable to their political constituency. The criticism often heard is that bureaucracy is heavy and efficiency low. But not only that – at times also attacks for information manipulation. For instance, the Concord network of NGOs writes on their 2006 Aid Watch Report:

> NGOs criticised key EU member states including the UK, France and Germany for inflating their aid figures. NGOs provide evidence that a total of €12.5 billion of headline EU aid in 2005 did not result in additional money for poverty reduction but was spent on debt cancellation, housing refugees and educating foreign students in European universities. In its briefing, the coalition of European organisations and national platforms representing hundreds of NGOs across Europe, called on EU governments to live up to their promises and called for new rules to ensure that debt cancellation does not come at the expense of new aid for developing countries. (Concord 2007)

The message is that the Poor are targeted in a mis-leading way since development aid has other supplementary agendas. Development aid also opens up budgets for hydropower project financing, functioning programmatically with reference to energy needs for development. In actual fact this aid might not be needed for other purposes than speeding up a lucrative investment process, seen in the eyes of hydropower companies. They have to make certain adjustments in political language, and they have to allocate some resources into side events. But in return they can speed up the already existing production plan and make money quicker.

The presentation of benefits from hydropower energy production by the International Hydropower Association starts off its account for social aspects with making references to supporting the poor:

> Hydropower schemes have the ability to significantly reduce poverty and enhance quality of life in the communities they serve. Access to electricity promotes new economic activity, empowers women by reducing domestic and repetitive chores such as firewood collection, improves health and education services, and provides a cleaner and healthier home environment. Hydropower infrastructure, such as reservoirs, also provides multiple-use benefits, particularly through increased availability, reliability and quality of fresh water supplies and reduced flood risks. (IHA 2004:17).

The quote is followed by sections on Managing social impacts, Outcomes for new developments, and Strategies to achieve proposed outcomes. The quote gives the impression that hydropower planning is linked into regional development plans, so that costs and benefits include calculations of the positive effects from poverty reduction and increasing numbers of enterprises, for instance. Chap. 3 has shown more of the reality. Hydropower schemes may contribute positively but over long periods of time. Furthermore, the schemes are commercial, based on classic BCR calculations, with sufficiently good return on invested capital. Reaching development aid resources, or loans from the equally commercial regional banks, such as ADB, will hardly influence more than priorities in a sequence of intended investment unless there is first an overall plan covering all potential water resources. Then radical priorities can be made, and other considerations included. The NHP Study seems to be about the only example in the world of such an approach. As already noted, it is then mistaken for hydropower project planning by critics who insist on taking it as contributing an EIA.

So the issue of poverty reduction can be seen to be treated in a programmatic way by two opposing sides also relating to hydropower development. An NGO side claims that whatever is done is too little, and is done in an ineffective way. Technically inclined hydropower developers on their turn claim that large-scale technology input will automatically reduce poverty if only managed properly. There is a case for mediation between these two positions, assuming an interest exists in finding compromise. However, it must be kept in mind that the language spoken is one of ideological statements. There is also a political language to be spoken, in line with the super-whale hypothesis.

5.4.2 The Free-floating Rivers – Whose Sustainability?

The super-whale syndrome is totemistic in the sense that people are classified as either for or against a group's icon. It identifies identity and prescribed ways of dealing with burning issues. Kalland (1998:9–13) attacks conservationists for being biased towards biodiversity issues, so the ecological systems shall be secured programmatically:

> Conservationists are primarily concerned with biodiversity. They promote balanced and sustainable utilization of natural resources. In other words, conservationists seek to secure ecological systems for future generations. The animal welfare movement is primarily concerned with the fate of individual animals. A broad range of nuances exist under this banner from those that demand more humane treatment and killing of animals, to the more radical element opposed to any use of animals whatsoever. This radical element promotes animal rights, and the most extreme example is the violent Animal Liberation Front which has been responsible for a number of bombings in the name of animal rights.

The point is made that seals and whales are not endangered species by scientific standards but treated as such by Greenpeace and WWF for legacy reasons. Business and industry can therefore buy a "green image". The parallel to hydropower seems to be in place: Selection of a vivid issue that has a dramatic illustration potential, falls within the environmental domain which worries many, and is available for campaign designing for lobbyist groups. In this respect it could be said that hydropower is like the super-whale. Only the icon becomes different, a free flowing river instead of a whale. And this becomes a super-river, not regulated, not polluted. In both cases the talk is about Nature not disturbed by civilization. Solidarity is firstly with Nature (international rivers or whales) and specific or complicating issues tend to be played down. Development like poverty reduction relating to hydropower sites, notably for ethnic minority situations, resettlement, and downstream issues, are not seen as development problems but as disturbances to Nature, and thus providing reasons for rejecting a hydropower project irrespective of analysis of other aspects of sustainable development. Mitigation issues become irrelevant since any project by definition is to be rejected. However, no realistic options for energy and related development are suggested to replace the "bad by definition" hydropower that may provide electricity and connected development inputs also for the rural poor. Rejections of this ideological position may also fall in the ideology sector; claiming that rejecting the South energy production is a colonial attitude that can lead to new apartheid thinking.

The Epupa project (see Chap. 2) provides a vivid example. It sparked off an environment conservation debate with roots in Namibia, but not in Angola. It was lifted to the international level very effectively by a string of organizations, both NGOs such as International Rivers Network, and donors such as the Swedish Sida, and also from inside the Namibian environmental ministry quarters at the time. The justification for paying much attention, indeed putting the project on the international agenda, was different in the various organizations – spanning from the NGO super-whale performance to the donor's efforts to avoid conflict by being transparent. Two different political languages were spoken, one rhetoric and one fact-finding. They frequently intertwined, as was shown in the "independent" assessment done by the International Rivers Network through a US University professor who gave his name to a document with frequent misunderstandings and without stakeholder discussion.

This kind of commentaries without verifications with project staff (also the SUAS study requested by Sida on the NHP Study mentioned above) can be approached as political language. Both discourses included a number of serious mistakes about the social issues dealt with. In both cases the consultants preferred to keep quiet to let the storm pass. This was felt to be the best political language. It was furthermore also in their best interest in another way; not to obstruct future potential contracts with their clients by being lured into a high-pitch debate on the international scene.

There is an element of uncertainty implicit in the new situation caused by the new environmental conventions. Professional competence is lacking but under build-up in the involved organizations. The political dimension of this is, that once published there is little way to win attention to the corrections. Attempts to clarify can on the political arena be interpreted as a guilty conscience, a measure that is suspicious and in its existence becomes another symptom on wrong-doing. The parallel to the study by Goffman (1959) on asylums is striking. There, no inmate, whether healthy or not, stood a chance to claim being healthy – efforts would just be registered as another disease symptom. Similarly, entering a debate may run the risk for a consultant or her/his client to be understood as defending a sensitive case.

5.4.3 The Failing Consultant, the Case of Uncritical Criticism

For sustainable development this kind of imbalance in the political debate needs to be countered towards constructive dialogue. The consultant is by critics felt to have the upper hand in a debate over a specific project through a more intimate project insight. The current study accounts for some of the methodologies applied within consultancies as one contribution towards dialogue on perspectives and good practice. The room for maneuver between political ideologies should become more transparent, especially when adhering to the political Millennium Development Goals (United Nations 2000). The social goals, not least poverty reduction, are more clearly set. A dialogue could concern what constitutes good science. Concepts, study focus and delimitation are all open for definitions at project levels – for instance: What is a poverty line? So is methodology – for instance: Is there an independent SIA or not? There is an uncertainty and vulnerability feeling that can open into a debate.

The cases studied contain a number of examples where the consultants, being the ones working in the field, have been able to identify suitable measures, approaches etc., but where initiative has been held back. When they could not be followed through, this becomes the target for the political ideological debate, not the fact that constructive efforts were made to improve a project. In a few instances the consultants have taken initiative "free of charge" (outside their Terms of Reference). One example is how the Baynes site option was added to the Epupa study, another the addition of a series of province/district stakeholders workshops in NHP when this proved feasible within budget constraints. In both cases evaluators have been asking for more activities instead of appreciating that initiative was taken.

The role as consultant easily becomes one of a middleman – the organization that takes all negative criticism. And it is paid for that, also in the own eyes. The point for this study is the effect – the consultant strives to stay within its terms of reference and refrains from creative thinking when new aspects open up in the light of an implementation process. And this becomes particularly destructive with the new approach to involve stakeholders, including local and regional persons, but also NGOs, more and more. Local knowledge and for the technical experts new experience may find difficulties to reach sufficient response more because of this context for them than because of what they wish to say.

5.4.4 Stakeholder Mobilization – A Political Force?

One of the main issues out of the hydropower projects discussed, covering the last decade, is that stakeholder mobilization has grown. This reflects a global trend to move up the stakeholder ladder, from the level of informing directly affected persons what is coming to them, to interacting over data formation first, then even over what data should be collected, and into mobilization into defining mitigation of negative consequences. The upgrading has not been self-evident since that one too, is a political process. Stakeholders are no longer passive objects but new players on the development scene. A new type of knowledge for hydropower development, local and regional experiences in particular, contributes to intensifying design, planning and implementation.

However, setting up a process of stakeholder participation involves learning and capacity building. The target sounds nice and radical, and is today part of the development nomenklatura. But it needs to involve informed stakeholder decision making. For hydropower the next step is to actually focus this process on the positive as well as the negative consequences of hydropower development would come natural also in implementation. The type of information dealt with in the current study would then become mandatory.

The new aspect is the political process. It is no longer sufficient in this tradition, to touch base on occasional workshops. Instead stakeholders get involved within a framework of rights and duties in a way that can not be fully controlled from outside. This might be the force that is needed to break up the complications indicated in last sections. In another era it would be peoples' revolution. Now it is stakeholder involvement. The process does not come automatically. It includes facilitation between stakeholder groups, between the different players, and also within a project between its different segments. In Vietnam at least, borders to technical and economic specialists are hard to penetrate for stakeholders interested in social issues. An unusual breakthrough, village population interacting with local government in an integrated water resource project, was taking place recently in Nam Puoi village and set an example for the entire province (Ngoc & Hjort-af-Ornas 2007). That study shows what is politically possible through decentralization, good governance, and extensive capacity building. It also demonstrates the complexity in this process. A number of specific issues have emerged and been dealt with step by step. The process has been time consuming and expensive. The end result is satisfaction among all actors; villagers as local stakeholders, local as well as regional and national administrators, and donor.

The example set in Nam Puoi can serve as an illustration. It is a stakeholder process that has emerged on local initiative. Even though it has been financed over separate budget (province and donor) there is a potential to afford the approach also in regular budgets, once the administration culture is in line with stakeholder involvement in a rights based tradition.

5.4.5 The Political Complications – Part of Hydropower Implementation?

Illustrations of the gap between implementation and policy can be found in several ongoing hydropower projects. The approach when applying policy is normally to follow safe-guarding guidelines on resettlement, ethnic minority (or indigenous people), livelihood, gender, and health/disease. A common thinking is that no one affected shall be worse off after the project implementation. Yet, differences in application also occur with respect to differences between the policy of a donor or bank, and that of a host country. The implementor may end up with several guidelines that must be negotiated to the best of her/his knowledge.

The guidelines have in many of the quoted projects been followed in a stereotyped manner and led to different interpretations; inside both the consultant's team and that of the client. And this in turn has triggered a debate over who has the final responsibility for a hydropower project. A formally correct answer to this issue is usually that project owners – governments – have to stand accountable for the result. A key issue becomes how to interpret new political targets, however, causing differences or changes in problem perceptions. Where the technical persons on both sides wish to follow a fixed, original work plan towards construction, the social persons protest that decisions can not be taken before technical baseline data can be presented to those directly affected. In the ethical turmoil that arises during hectic times of meeting agreed deadlines, the newly upgraded rights and responsibilities of the PAPs have largely disappeared from the agenda in times of project delivery stress, judging from the results in all the quoted cases, such as in Song Bung 4. While project leaderships remain with the normal management risk under uncertainties, the new demands for impact assessments have been seen as bringing further complications in stress situations; that is essentially when trying to keep a time plan. In those instances the agenda to save a project and its key players takes over from meeting "soft" policy demands on the social side. In certain instances guidelines were in the end simplified or not even applied by the client itself when staff became busy to save the project's time plan.

Such a situation is possible to avoid if it has been prepared for in early stages of project design. There are discrepancies to resolve, for instance between ADB and

Vietnam on definitions of indirectly affected, or even differences between provinces, such as in the decentralized Vietnam. Differences need to be handled before implementation; there is no short-cut. Considerations of the political and economic implications of such agreements, for instance the compensation to downstream populations, leads to politically charged issues. Claims can be made in retrospect for already existing projects if agreements are reached about compensation rates. This is an underlying hidden agenda in the current (2007) negotiations between Cambodia and Vietnam concerning the consequences caused by hydropower development, for instance.

5.5 The New Hydropower Project Culture

This study has painted the picture of how hydropower project implementation, and later monitoring, have followed suit with the global environmental conventions. It has showed how old issues in the periphery of studies have been upgraded, and the techniques to achieve this. It has demonstrated delays and gradual adaptations in pace with a learning process and changes in problem perceptions among professionals, mostly technically trained people. It has also touched on the communication of a conservationist ideology that does not accept large-scale hydropower energy production neither as renewable nor as environment friendly. Some concerns can be raised over the emergence of a new hydropower culture paradigm, but that is outside this study. The focus here is on lessons learnt for sustainable development implementation from one selected case; hydropower. And the observation is that awareness raising is needed among the actors within hydropower, so that they can tune in with the already agreed requirements.

This issue of integration between what at present seems to be competing ideological positions is one area where the current study can contribute with some insight. It offers the experience from some significant hydropower projects so that the tension between policy intention and implementation shows. This vibration is a healthy sign, given the expectation that the global environment conventions should bring about change for hydropower. The Millennium Development Goals, centered on doing away with acute poverty before 2015 as one condition to sustainable development, call for more than cosmetics. By addressing the social issue in its own right and not merely as an aspect of natural resource relations, the study intends to learn from past experiences in order to widen the scope from social safe-guarding (ADB and WB concept), to better target poverty reduction and upgrade stakeholder involvement.

The issues raised in the current study have indicated how policy-related concerns within hydropower that are significant also broader for sustainable development thinking surface in project implementation. They relate to the core problem: How the new sustainable development thinking can gain its foothold also in large-scale and technically dominated projects, such as hydropower production. This is an issue that includes changes in style but also in outlook. It seems to be a common concern in large-scale projects. Hydropower provides a conspicuous case, with its deep environmental and social consequences along with the infrastructure significance. Special attention is needed to issues of political power relations. The actors and their agendas are found in the context of technical, economic, environmental and social considerations for hydropower. The current study is particularly concentrated on the social issues. The search has been for the implementation patterns based on policy derived from the global environmental conventions. The emergence of strengthened sustainability targets can then be followed up by action, primarily by those who are affected regionally (sub-national) and locally. A rights perspective, highlighting both rights and obligations, seems to fit the needs in a good way. Such perspective will bring about a shift in paradigm. The new management of fact finding will be assessed in terms of administrations' and consulting/implementing companies' adjustments to new winds of change.

References

ADB 2004, *Effectiveness of Participatory Approaches: Do the New Approaches Offer an Effective Solution to the Conventional Problems in Rural Development Projects?* Operations Evaluation Department. SST: REG 2005-01. December.

ADB 2005a, *2nd Red River Basin Sector Project, Part A: Water Resources Management*, Draft final report TA 3892. Asian Development Bank and Ministry of Agriculture and Rural Development Vietnam. Sweco/Groner & Delft Hydraulics. October.

ADB 2005b, *2nd Red River Basin Sector Project, Part A, Phase 2. Component 3*. Draft final report TA 3892. Asian Development Bank and Ministry of Agriculture and Rural Development Vietnam. Sweco/Groner & Delft Hydraulics. October.

Cambodia, Government of 2001, *Participatory Poverty Assessment in Cambodia*. Asian Development Bank.

Concord 2007, Network of European NEOs. http://www.bond.org.uk

CPRGS 2002, *Comprehensive Poverty Reduction and Growth Strategy*. Socialist Republic of Vietnam. Document No. 2685/VPCP-QHQT, May.

Dedolph, C. 2002, Mainstreaming Participation in ADB, *ADB Review* March–April.

Garcia, J. C., Garcia, C. T., Devkota, S., & Thanju, R. P. 2005, Resettlement: Lessons learned at Kali Gandaki A in Nepal *HRW*:33–37. March.

Goffman, I. 1959, *Asylums*. Penguin.

Granfelt, T., & Hjort-af-Ornas, A. 2003, *Näringslivsklimatet i kommunerna: En fråga om kultur och samspel* [Enterprise climate in the communes: A matter of culture and interaction] Centrum för kommunstrategiska studier, Linköpings universitet. Rapport 2003:3.

Harper, M. 1984, *Small businesses in the Third World. Guidelines for Practical Assistance.* ITDG Publications, London.

Hjort, A. 1974, *Socio-economic Effects of Rural Electrification in Kenya*. Stockholm: University of Stockholm, Department of Social Anthropology.

Hjort-af-Ornas, A., & Ngoc, P. T. B. 2004, *Water Sub-sectors and the Poor in Northern Vietnam*. Paper for Vietnam/Sweden research cooperation workshop, Hanoi, December. Sida.

Hjort-af-Ornas, A., & Ngoc, P. T. B. 2007, *Landslide, resettlement and stakeholder performance in Nam Puoi*, Sida, Hanoi.

HydroConsult 2006, *Modelling Low Flow Effects from Large dams*. Travel report by Hjort-af-Ornas, A. Costa Rica Electricity Board.

Instituto Nicaraguense de Energia/Norconsult International/SwedPower 1993, Environmental and Socio Economic Impact Assessment of the Los Calpules Hydropower Project, Proyecto Rio Viejo, Los Calpules, Nicaragua.

Instituto Nicaraguense de Energia/Norconsult International/SwedPower 1994, Environmental Impact Assessment of the La Sirena hydropower project, NicaraguaFeasibility study.

International Energy Agency 2002, *World Energy Outlook*. OECD. November.

International Hydropower Association 2004, *Sustainability Guidelines*. February.

Kalland, A. 1998, The superwhale: Hijacking myth and symbol in environmentalism. In *DARK NIGHT Field Notes*. See http://dbh.nsd.ub.no/nfi/litteratur (An internet journal) Vol. 14, pp. 9–13.

Kenya, Government of 1996, Kenya: Participatory poverty assessment. *Environmentally and Socially Sustainable Development Network*. Note No. 26, World Bank. August.

Komives, K., Foster, V., Halpern, J, & Wodon, Q. 2005, With support from Abdulah, R. *Water, Electricity, and the Poor. Who Benefits from Utility Subsidies?* Directions in development. The World Bank. Washington.

Kugbei, S., & Turner, M. 2000, Structure and establishment of small-scale enterprises in seed industries of developing countries. In Kugbei, S. et al. (eds.) pp. 9–14.

Lalander, B. 1971, *Rural Electrification in Kenya*. Sida.

Lao PDR 2001, *Participatory Poverty Assessment, Attapeu Province*. Actionaid International.

Lindskog, E., & Long, V. N. 2004, *Resettled but not Restored. Evaluation of the Resettlement Process in Song Hinh Hydropower Multi-purpose Project, Phu Yen Province, Vietnam*. Institute of Tropical Biology in Ho Chi Minh City & SEI, Stockholm Environment Institute. January.

Marcelle, G. M., & Jacob, M. 1995, Chapter 11, The "double blind". In *Missing Links. Gender Equity in Science and Technology for Development*. Gender Working Group of the United Nations Commission on Science and Technology for Development IDRC/ITDG Publishing/UNIFEM.

Mathews, C., & Chesselet, J. 1997, *Cunene Film Project*. Film 3. Doxa Production.

MDG 2000, *Millennium Development Goals*, Report. United Nations, New York.

Narayan, D., Patel, R., Schafft, K., Rademacher, A., & Koch-Schulte, S. 2000, *Voices from the Poor. Can Anyone hear us?* World Bank 2000: World Bank, Oxford University Press.

Ngoc, P.T.B. & Hjort-af-Ornas A. 2007, A Village Fearing Land Slide. Lessons from the voluntary resettlement of Nam Puoi. *Newsletter Natural Disaster Mitigation Partnership*, Vol. 4.

Nicaragua 2000, *Assessment of Hydroelectric Alternatives*. SWECO and INE.

Norplan, Namang Concortium 1996, *Epupa Hydropower Scheme, Feasibility Study, Project Formulation Report, Part III, Comparative Environmental Assessment*. NamAng Consortium of Consulting Engineers for the Feasibility Study of the Epupa Hydropower Scheme.

Orgut 2002, *Thematic Evaluation of Rural Electrification Projects*. Stockholm.

Pakistan, Government of 2003, *Pakistan Participatory Poverty Assessment. Balochistan Province Report*. Asian Development Bank.

Sechaba Consultants 2000, *Poverty and livelihoods in Lesotho, 2000*. More than a mapping exercise. Maseru. June.

Sen, A. 1981, *Poverty and Famines*. Oxford: Clarendon Press.

Sida 2004, *Civil Society and Poverty Reduction*. A guide for development practitioners. J. Manor. Stockholm.

Sida 2005, *Sustainable Energy Services. Policy for Poverty Reduction*. December.

Soussan, J. 2003, *Water and Poverty. Fighting Poverty Through Water Management*. ADB.

Sullivan, C. 2001, The potential for calculating a meaningful water poverty index. *Water International* 26(4), 471–480.

SWECO 1998, *National Hydropower Plan Study, Stage 1*. SWECO, Statkraft and Norplan.

SWECO 2001, *Song Hinh Multipurpose Project*. Hanoi.

SWECO 2004, *National Hydropower Plan Study, Stage 2*. SWECO, Statkraft and Norplan.

SWECO 2006, *Song Bung 4 Hydropower Project, Phase II. Draft Final Report. Resettlement and ethnic minority development plan. Volume 1: Cross cutting issues*. TA 4625-VIE. Asian Development Bank. October.

Swedish Waterhouse 2005, Future dams. Recommendations to Swedish stakeholders on implementing "Dams and development – A new framework for decision making". Swedish Committee for Water and Dam Issues (SKYD), Stockholm.

UNCED 1992, United Nations Conference on Environment and Development. New York: Oxford University Press.

UNESCO 2006, *The Global Network of Water Anthropology (NETWA)*. International Hydrology Program. Paris. March.

United Nations 2000, *Millennium Development Goals*. Millennium Declaration.

WCED 1987, *United Nations World Commission on the Environment and Development, Our Common Future*. Oxford: Oxford University Press.

Vietnam Development Report 2004, *Poverty*. Joint Donor Report to the Vietnam Consultative Group Meeting, Hanoi, December 2–3, 2003.

Vietnam News 2006, New Poverty Line. January 3.

Vietnam, Socialist Government of 2002, *The Comprehensive Poverty Reduction and Growth Strategy (CPRGS)*. Hanoi, May.

Wallentinus, H-G. et al. 2005, *Monitoring of Environmental and Social Aspects in the National Hydropower Plan (NHP) Study in Vietnam*. Sida Help Desk, SLU.

Wason, D., & Hall, D. 2002, *An Overview of Household Economic Status and Government Policy*. CPRC Working Paper No. 40.

World Commission on Dams 2000, *Dams and Development. A New Framework for Decision-Making*.

World Wildlife Foundation 2005, *To Dam or not to Dam? Five Years on from the World Commission on Dams*. Dam right. WWF's dams initiative.

WSSD 2002, *World Summit on Sustainable Development*. United Nations.

Printing: Krips bv, Meppel, The Netherlands
Binding: Stürtz, Würzburg, Germany